Floods
Hydrological, Sedimentological and Geomorphological Implications

Edited by
Keith Beven
Centre for Research on Environmental Systems
University of Lancaster

and

Paul Carling
Freshwater Biological Association
Far Sawrey, Ambleside, Cumbria

John Wiley and Sons
CHICHESTER · NEW YORK · BRISBANE · TORONTO · SINGAPORE

Copyright © 1989 by John Wiley & Sons Ltd.

All rights reserved.

No part of this book may be reproduced by any means, or transmitted, or translated into machine language without the written permission of the publisher.

Library of Congress Cataloging-in-Publication Data:
Floods: hydrological, sedimentological, and geomorphological implications/edited by Keith Beven and Paul Carling.
  p. cm. — (British Geomorphological Research Group symposia series).
  Bibliography: p.
  Includes index.
  ISBN 0 471 92164 5
  1. Floods—Congresses.   I. Beven, Keith.   II. Carling, Paul.
  III. Series.
  GB 1399.F59 1989
  551.48′9—dc 19

89-5311
CIP

British Library Cataloguing in Publication Data
Floods.
  1. Floods: their hydrological, sedimentological and geomorpholocial implications
  I. Beven, Keith   II. Carling, Paul   III. Series
  551.48′9

ISBN 0 471 92164 5

Printed and bound in Great Britain at the Bath Press, Bath, Avon

Contents

vii *Series Preface*

ix *Preface*

xii *List of Contributors*

1 1 The Hydrology, Sedimentology and Geomorphological Implications of Floods: an Overview
   P. CARLING and K. BEVEN
11 2 Storm Runoff Generation in Small Catchments in Relation to the Flood Response of Large Basins
   T. P. BURT
37 3 Flood Wave Attenuation due to Channel and Floodplain Storage and Effects on Flood Frequency
   D. R. ARCHER
47 4 Physically Based Hydrological Models for Flood Computations
   S. AMBRUS, L. IRITZ and A. SZÖLLÖSI-NAGY
57 5 Flood Frequency and Urban-Induced Channel Change: Some British Examples
   C. R. ROBERTS
83 6 Hydraulics of Flood Channels
   D. W. KNIGHT
107 7 Flow-Competence Evaluations of the Hydraulic Parameters of Floods: an Assessment of the Technique
   P. D. KOMAR
135 8 Floods and Flood Sediments at River Confluences
   I. REID, J. L. BEST and L. E. FROSTICK
151 9 Flood Effectiveness in River Basins: Progress in Britain in a Decade of Drought
   M. D. NEWSON
171 10 Magnitude and Frequency of Palaeofloods
   V. R. BAKER
185 11 The Use of Soil Information in the Assessment of the Incidence and Magnitude of Historical Flood Events in Upland Britain
   R. F. SMITH and J. BOARDMAN

199	12	The Yellow River (County Leitrim, Ireland) Flash Flood of June 1986 P. COXON, C. E. COXON and R. H. THORN
219	13	River Channel Changes in Response to Flooding in the Upper River Dee Catchment, Aberdeenshire, over the Last 200 Years L. J. McEWEN
239	14	Sedimentology and Palaeohydrology of Holocene Flood Deposits in Front of a Jökulhlaup Glacier, South Iceland J. MAIZELS
253	15	Flood Deposits Present within the Severn Main Terrace M. DAWSON
265	16	Floods in Fluvial Geomorphology J. LEWIN

Series Preface

The British Geomorphological Research Group (BGRG) is a national multi-disciplinary society whose object is 'the advancement of research and education in Geomorphology'. Today, the BGRG enjoys an international reputation and has a strong membership from both Britain and overseas. Indeed, the Group has been actively involved in stimulating the development of geomorphology and geomorphological societies in several countries. The BGRG was constituted in 1961 but its beginnings lie in a meeting held in Sheffield under the chairmanship of Professor D. L. Linton in 1958. Throughout its development the Group has sustained important links with both the Institute of British Geographers and the Geological Society of London.

Over the past three decades the BGRG has been highly successful and productive. This is reflected not least by BGRG publications. Following its launch in 1976 the Group's journal, *Earth Surface Processes* (since 1981 *Earth Surface Process and Landforms*) has become acclaimed internationally as a leader in its field, and to a large extent the Journal has been responsible for advancing the reputation of the BGRG. In addition to an impressive list of other publications on technical and educational issues, BGRG symposia have led to the production of a number of important works including *Nearshore Sediment Dynamics and Sedimentation* edited by J. R. Hails and A. P. Carr; *Geomorphology and Climate* edited by E. Derbyshire; *River Channel Changes* edited by E. Derbyshire; *River Channel Changes* edited by K. J. Gregory; and *Timescales in Geomorphology* edited by R. Cullingford, D. Davidson and J. Lewin. This sequence of books culminated in 1987 with a publication of the *Proceedings of the First International Geomorphology Conference* edited by Vince Gardiner. This international meeting, arguably the most important in the history of geomorphology, provided the foundation for the development of geomorphology into the next century.

This open-ended BGRG Symposia Series has been founded and is now being fostered to help maintain the research momentum generated during the past three decades, as well as to further the widening of knowledge in component fields of geomorphological endeavour. The series consists of authoritative volumes based on the themes of BGRG meetings, incorporating, where appropriate, invited contributions to complement chapters selected from presentations at these meetings under the guidance and editorship of one or more suitable specialists. Whilst maintaining a strong emphasis on pure geomorphological research, BGRG meetings are diversifying, in a very positive way, to consider links between geomorphology *per se* and other disciplines such as ecology, agriculture, engineering and planning. The first volume in the series on *Geomorphology in Environmental Planning*, edited by Janet Hooke, exemplifies this new trend. This volume, the second in the series, returns to a traditional theme, namely on the role of extreme events in landscape evolution. Floods—Hydrological, Sedimentological and Geomorphological Implications edited by Keith Beven and Paul Carling, provides a state-of-the-art review and reflects the continuing buoyancy of fluvial geomorphology internationally.

The BGRG Symposia Series will contribute to advancing geomorphological research and we look forward to the effective participation of geomorphologists and other scientists concerned with earth surface processes and landforms, their relation to Man, and their interaction with the other components of the Biosphere.

14 October 1988

Geoffrey Petts
BGRG Publications

Preface

This volume of papers arises from a workshop held at the University of Lancaster as a joint meeting of the British Geomorphological Research Group and the British Hydrological Society. The main aim of the workshop was to bring together scientific researchers with a common interest in the dynamics of fluvial floods and their effects on the landscape. Floods are the extremes of the hydrological record, events of large magnitude that occur only rarely. They are of interest to government authorities, water supply and waste disposal bodies, the emergency services, insurance agencies, and others. Planning for and dealing with the effects of flood events can be facilitated by greater scientific understanding of the mechanisms involved, but this has been the concern of scientists from a variety of different disciplines, such as hydrologists, hydraulic engineers, sedimentologists, and geomorphologists. The Lancaster workshop was intended to provide a forum for presentations and discussion involving all these areas of expertise and to encourage the sharing of experience and understanding between the different disciplines.

We would like to thank everyone who contributed in making the workshop a success, in particular the support of the British Geomorphological Research Group, British Hydrological Society, Pennines Hydrology Group, the Royal Society of London, the University of Lancaster and the Freshwater Biological Association. The wide range of material presented and discussed during the three days was a tribute to the speakers, the session chairmen and lady, and the audience alike. We would particularly like to thank Adrian Harvey of the University of Liverpool who led a very successful field trip to examine the effects of flood events in the valleys of the Howgill Fells in Cumbria. Final thanks are due to the many referees of the papers included in this volume who helped shape its final form.

June 1988

Keith Beven
Paul Carling

List of Contributors

Sándor Ambrus, Water Resources Research Centre, POB 27, H-1453 Budapest, Hungary.

David R. Archer, Northumbrian Water.

Victor R. Baker, Department of Geosciences, University of Arizona, Tucson, Arizona 85721, USA.

James L. Best, Department of Geology, University of Leeds, Leeds, LS2 9JT.

Keith Beven, Institute of Environmental and Biological Sciences, University of Lancaster, Bailrigg, Lancaster, LA1 4YQ.

John Boardman, Countryside Research Unit, Brighton Polytechnic, Falmer, Brighton, BN1 9PH.

Tim P. Burt, School of Geography, University of Oxford, Mansfield Road, Oxford, OX1 3TB.

Paul Carling, Freshwater Biological Association, Far Sawrey, Ambleside, Cumbria.

P. Coxon, Department of Geography, Trinity College Dublin 2, Ireland.

C. E. Coxon, Environmental Sciences Unit, Trinity College Dublin 2, Ireland.

Martin Dawson, BP International Ltd, Britannic House, Moor Lane, London, EC2Y 9BU.

Lynne E. Frostick, Department of Geology, Royal Holloway and Bedford New College, Egham Hill, Egham, Surrey, TW20 0EX.

László Iritz, Uppsala University, Division of Hydrology, V. Ågatan 24, S-752 20 Uppsala, Sweden.

Donald W. Knight, Civil Engineering Department, University of Birmingham, PO Box 363, Birmingham, B15 2TT.

Paul D. Komar, College of Oceanography, Oregon State University, Corvallis, Oregon 97331, USA.

John Lewin, Department of Geography, University College of Wales, Aberystwyth, Llandinam Building, Penglais, Aberystwyth, SY23 3DB.

Judith Maizels, Department of Geography, University of Aberdeen, St Mary's High Street, Old Aberdeen, AB9 2UF.

Lindsey J. McEwen, Department of Geography and Geology, The College of St Paul and St Mary, The Park, Cheltenham, Gloucestershire, GL50 2RH.

Malcolm D. Newson, Department of Geography, University of Newcastle-upon-Tyne, Newcastle-upon-Tyne, NE1 7RU.

Ian Reid, Department of Geography, Birkbeck College, 7–15 Gresse Street, London, W1P 1PA.

Carolyn R. Roberts, School of Geography and Geology, College of St Paul and St Mary, The Park, Cheltenham, Gloucestershire, GL50 2RH.

R. F. Smith, Countryside Research Unit, Brighton Polytechnic, Falmer, Brighton, BN1 9PH.

András Szöllösi-Nagy, Water Resources Research Centre, POB 27, H-1453 Budapest, Hungary.

R. H. Thorn, School of Science, Sligo Regional Technical College, Sligo, Ireland.

Floods: Hydrological, Sedimentological and Geomorphological Implications
Edited by K. Beven and P. Carling
© 1989 John Wiley & Sons Ltd

1 The Hydrology, Sedimentology and Geomorphological Implications of Floods: an Overview

PAUL CARLING
Freshwater Biological Association, Far Sawrey, Ambleside

KEITH BEVEN
Institute of Environmental and Biological Sciences, University of Lancaster, Lancaster

> I am not ſo much a Philoſopher, as to find out what could occaſion ſuch a vaſt Collection of Clouds or Vapours, particularly at that Time and Place, but am ſatisfy'd, from the Havock it has made in ſo ſhort a time (for it was all over in leſs than two Hours), that it muſt have far exceeded any Thunder-ſhower that we have ever ſeen: moſt probably a Spout, or large Body of Water, which, by the rarefaction of the Air occaſioned by that inceſſant Lightning, broke all at once upon the Tops of theſe Mountains, and ſo came down in a Sheet of Water upon the Valley below.
>
> Clarke, 1749

> Theſe Thunder-clouds, with equal force and contrary Directions, met together upon the Mountains above the Valleys of St. Johns And Threlkeld, as at or about the great Dod and Cova pike, and muſt of conſequence hover on or about them, and thereon vent Water-ſpouts, (but not ſo on the Valleys, otherwiſe than by the violent Courſe of the Brooks and Rivulets from the one down to the other); which would increaſe and perpetuate the Lightning, ſo ſwift in Motion, and viſible to our Eyes, but retard and obſtruct the Undulations of the Air, which are far more ſlow in Motion, and later in coming to our Ears; for any two ſuch Bodies as thick Clouds, driven by contrary Winds, and meeting together with equal Force and contrary Directions, cannot impel each other backwards or forwards, but muſt remain at or about the Place where they met, and there exert their Vigour; which, in this Caſe,

muſt be the Reaſon of ſuch Water-ſpouts upon theſe Mountains, and not in the Valleys.

Naughley, 1749

It is over two hundred years since Clarke and Naughley speculated on the causes and consequences of thunderstorm activity in the Lakeland Hills above the Vale of St John (Hutchinson, 1749). Today, their speculations, circumscribed by uncertainties, may seem somewhat insubstantial, but careful reading of the various contemporary accounts indicates a nascent awareness and growing interest in the role forces of nature play in shaping the landscape. This early attempt at developing scientific method, linking forcing function, process, and resultant form is even the more remarkable given that contemporary explanations for these floods included effluxes of water from the bowels of the earth induced by volcanic activity!

Informed fundamental arguments of the time were summarised a few years later by G. Smith (1754) in diagrammatic form (Frontispiece). The highest hills (F) intensified the storm through orographic uplift and runoff was concentrated over the slopes of the lower hills (H). Gullies were deepened (Q) by the floodwaters and the eroded material deposited as fans (E). Smith, in his text, made comments with respect to the eroding and transporting power of the streams and the destructive nature of streamflow, demolishing walls (G) and buildings (A, B), whilst some areas on the alluvial flat remained clear of the floodwaters (C). Smith explained the spatial variation in power in terms that today might be explained more familiarly through Duboy's and continuity equations and the streampower concept:

The hollows, or channels which wind down the declivity, and when I ſaw them were dry, become gradually wider, and more ſhallow, as they deſcend to that part of the montain which is leſs ſteep; the waters in proportion as they ſpread, loſe their force, cover a larger tract, and fall with leſs rapidity

Such accounts indicate that, although process–response models were poorly defined in the eighteenth century, the observers realised that any explanatory models must be spatially and temporally complex, linking atmospheric/topographic interactions, hydrological runoff processes and basin-scale factors, to produce elementary erosional and depositional forms varying in character depending upon their location within a catchment. In addition, the role of seemingly catastrophic events must needs be integrated to a lesser or greater extent with understanding of processes observable over shorter time spans, such as the day-to-day and seasonal experiences of the agricultural natives of Lakeland. To this end, the scientific gentlemen of the eighteenth century interviewed local inhabitants to secure a perspective on what might be considered 'usual' or 'catastrophic' in the Lake District environment.

Hydrology, geomorphology and sedimentology have advanced a lot since the eighteenth century. Even today, however, it is not too clear what makes a flood different from other events. Our concepts of flood frequency lead us to assume that all events are drawn from some underlying probability distribution. We may disagree about the form of that distribution—log normal, Gumbel, two-component exponential, EV2, EV3, log Pearson type III, Wakeby or Weibull—and particularly of the upper tail of extreme values which we rarely have sufficient data to determine with any accuracy, but the assumption is that the distribution is continuous. Thus there is no obvious threshold for a flood event in the discharge record. Some other criterion needs to be invoked, such as the occurrence of overbank flows, except that in many channels of irregular cross section it may be very difficult to define such a physical limit. Our textbooks tell us that bankfull discharge, of course, occurs on average with a return period of about 1.5 years (Wolman and Leopold, 1957; Williams, 1978), and this may be a suitable working definition of a flood flow. But what makes a flood different from other events? Is it differences in runoff production; differences in sediment transport processes; differences in erosional and depositional process; or just differences in the magnitudes of the effects relative to smaller events?

It seems that it is difficult to give a general answer to this question. The study of flood events inevitably involves probabilistic considerations of magnitude and frequency in relation to the effects of a particular flood event. The *effectiveness* of an event of a certain magnitude has proved difficult to define. In terms of the human and economic impacts of flooding, peak magnitude or stage may be of primary importance, but it is clear that as far as the sedimentological and geomorphological effects of floods are concerned the relationship between magnitude and effect may be much more complex. Early concepts of effectiveness in the geomorphological literature were centred around the ideas of Wolman and Miller (1960):

> The effectiveness of processes which control many land forms depends upon their distribution in time as well as their magnitude. It cannot be assumed that simply because of their magnitude the rare or infrequent event must be the most effective. Analyses of the transport of sediment by various media indicate that a major portion of the work is performed by events of moderate magnitude which recur relatively frequently rather than by rare events of unusual magnitude. (p. 70)

The availability of sediment and sequence of events may be as important as magnitude in determining the perceived geomorphological effects on hillslopes and channels (Newson, 1980; Beven, 1981). It is clear that many rare flood events are perceived as having very little geomorphololological effect, especially in lowland regions (Dury, 1973), while in other situations environmental change may be dominated by the effects of rare catastrophic events (see discussion of Wolman and Gerson, 1978).

It may be useful to reconsider the effects of floods in the light of recent research activity in the area of nonlinear dynamic systems (see for example Prigogine and Stengers, 1984; Huggett, 1988). These open hydrological and geomorphological systems are subject to continuously varying stochastic inputs and develop through time, sometimes in an apparently chaotic matter (such as the historical variations in the width of the Gila River, see Stephens *et al.*, 1975). However, they are also subject to the constraints imposed by the conservation of mass and energy and to the physical constraints of material properties. The result is a hierarchy of order within the system that may involve time variable thresholds, such as the position of channel heads and flood plain initiation. The historical context is important in determining the form of that organisation, particularly in geomorphological systems where the time scales of hillslope and channel development may be long relative to non-stationarity in the climatic regime of the inputs. The concept of relaxation time (see for example Anderson and Calver, 1977) is important in this context, since in geomorphological systems there may be a hierarchy of relaxation times associated with different geomorphological features, gradually being modified by sequences of events of different magnitudes.

Within this framework it may be difficult to define the geomorphological effectiveness of an event of a given magnitude, particularly if we have only a short-term record of observations, as is usually the case. Effects that may occur associated with one particular flood event may be absent after a later event of the same magnitude or larger, if the time between the events is shorter than the relaxation time associated with the effects of the first event. That relaxation time may itself depend upon the sequence of intervening events. This view of the effectiveness of floods is one of complexity, and the studies contained in this volume represent attempts to grapple with this complexity. In this there remains a common theme with our forebears of the eighteenth century.

In assessing the research effort of the last ten years, Newson (this volume, Chapter 9) emphasises that despite a decade characterised by drought periods, paradoxically we have extended our knowledge of the geomorphological effectiveness of floods, in as much as we are now aware of the complexities of possible modes of response in the role of extreme process rates and how these couple with more moderate rates. This interest is particularly timely given that a contraction of the hydrological measurement network is occurring at the same time as concern is growing that the so-called 'greenhouse effect' may significantly affect the UK hydrological balance by the end of the century.

We are at an interesting stage in the development of the environmental sciences that include hydrology, geomorphology and sedimentology, marked by a transition from studies that are aimed primarily at *understanding* of processes and their interactions within the landscape, to the first attempts at *quantitative prediction* of the effects of those processes. The majority of the papers in this volume are primarily concerned with the former, using analyses of controlled laboratory experiments (Knight), dating techniques (Baker), historical data (Roberts, McEwen), or the interpretation of field data (Archer, Reid *et al.*, Maizels, Coxon *et al*, Dawson, Komar) as a means of throwing light on the processes operating under extreme flow conditions. Some of these studies, such as the hydraulic relationships of Knight, and the regression relationships of Komar and Roberts might also be used at predictive tools, but that is not, as yet, the primary purpose of such studies.

Assessing the impact of future natural and anthropogenic perturbations to the environment requires predictive models that are physically realistic. The progress along the road to attaining predictive capability in these sciences is inevitably dependent upon the time scales of the responses. There is now over twenty years' experience of using digital computers for short-period hydrological forecasting (summarised for example in Anderson and Burt, 1986). Hydrological predictions of flood events are now made routinely for both gauged and ungauged sites, although the uncertainties in those predictions are not often assessed. Some short time scale geomorphological processes have also been successfully modelled, such as river sediment dynamics (e.g. Ponce *et al.*, 1979) Borah *et al.*, 1982; Lyn, 1987). However, applications of predictive models, to other than hypothetical situations, for the longer time scales involved in sedimentology and geomorphology have been relatively rare.

It is worth considering the prospects for developing a predictive capability in these areas. What remains to be done before predictive models can be made available? Consider first the case of hydrological models. Ambrus *et al.* (this volume, Chapter 4), show how it is possible to develop an operational predictive capability for short-term flood dynamics without the need for a full understanding of the physical processes involved. However, for predictions involving links between water flows and sediment transport processes, such black-box models will generally not be adequate. It may be important to predict the nature of both hillslope and channel-flow processes to properly characterise the nature of sediment transfers within a catchment. Over longer periods of time, the ways in which hillslope and channel-flow processes change with event magnitudes and sequences of events will be important in predicting geomorphological responses. Some of the complexities of the interactions between hydrological responses and sediment transport are summarised in the paper by Burt (this volume, Chapter 2).

Current physically based models in hydrology, such as the SHE model (Bathurst, 1986) or IHDM (Beven *et al.*, 1987, Calver, 1988) have developed to the point where distributed predictions can be made of surface and subsurface runoff production within a catchment. Water flows alone are predicted and the physical characteristics of a catchment are assumed to be virtually fixed in time, although components to predict short-term flood event sediment and solute transfers within the framework of the SHE model are currently under development (see Storm *et al.*, 1987). Clearly, the flow of water is the driving force for solute and sediment transfers and accuracy in the prediction of the latter must depend on the accuracy of predictions of water flows. It has been argued recently by Beven (1987a) that the current generation of physically based models of water flow may be inadequate in both conception and implementation as a result of spatial and temporal variability of catchment parameters, and problems in calibrating those parameters, albeit that they are 'physically based'. These analyses suggest that the uncertainties associated with current physically based hydrological predictions may be large, even for short-period flood event simulations, and point to the importance of additional field data in constraining those uncertainties.

The first steps are also being made in attempts to predict flood frequency characteristics for individual catchments from physical principles. A workable model of hydrological responses over a range of storm conditions of increasing severity must be a precursor to any long time scale geomorphological model. The flood frequency predictions require a number of different components, including a model for the occurrence of rainfall intensities in time and space for different storms, a model for runoff production, a model for runoff routing and a model for the interstorm periods to set up the antecedent conditions prior to the next runoff-producing storm event. To date most attempts at such flood frequency predictions have been based on the simplest of hydrological models, essentially a point infiltration model extended to the catchment scale by simple multiplication by a catchment or effective contributing area (e.g. Hebson and Wood, 1982; Cordova and Rodriguez-Iturbe, 1983; Diaz-Granados *et al.*, 1984). These models have not been too successful (see Moughamian *et al.*, 1987) and this must be due, at least in part, to inadequate characterisation of the effects of geomorphology on the runoff production process. The flood frequency model of Beven (1987b) does consider the spatial variability of topography in controlling runoff production contributing areas, and the way in which they change with antecedent conditions and storm severity but, as yet, it has been tested only for the Wye catchment. It does not consider any interactions between hydrological response, sediment transfers and geomorphological development even for the severest floods.

It would appear, therefore, that more understanding of hydrological processes will be required before an adequate predictive capability for geomorphological systems can be developed to cope with a variety of environments, basin scales, process thresholds and sequences of extreme events. Regional storms or other environmental changes covering entire basins ($>$ 100 km^2) may be required to induce cataclysmic changes (*sensu* Benson, 1984) 'down-on-the-plain', whilst localised convective storms affecting headwater basins ($<$ 10 km^2), where streampower is most effective (Baker and Costa, 1987), may be locally catastrophic, leading to short-lived restructuring rather than basin-wide long-term changes in the nature of the system. A suitable model must be able to handle the processes and thresholds involved, and make use of the information available from the interpretation of the field evidence.

For any major advance to be made in this direction it will be necessary to couple modelling studies with field studies in a synergistic way that may shed light, both on the nature of change in geomorphological systems and the ways in which hydrological processes reflect such changes. In the post-glacial period, both climate and land-use patterns have been changing, and these changes have been recorded in the terraces and flood-plain sediments of the large rivers. Even in such a populous country as the UK there is surprisingly little information about channel activity rates over historical time (let alone throughout the Holocene) and how channels have adjusted to changing flood frequency. It is evident from the study of McEwen (this volume, Chapter 13) for example that process thresholds are clearly involved in major changes in river plan form, and that the threshold mechanisms associated with 'avulsion' in particular have received inadequate attention.

A number of the papers in this volume stress that proper attention should be paid to the interpretation of flood sediments. Geomorphologists have generally paid insufficient attention to modern advances in sedimentology. What are the facies arrangements one might associate with a very large flood discharge in a braided river, or at a former junction with a major tributary? Are the processes operating today a key to those operating throughout the immediate post-glacial period? A few years ago, contemporary debris-flow activity was not a process recognised in the British uplands but this has been shown to be incorrect at least in smaller headwater catchments (Wells and Harvey, 1987; Carling, 1987). When our larger rivers were fed by receding glacial ice-sheets, till-flows and other debris-flows from vast tracts of unvegetated material, perhaps triggered by rapidly varying meltwater regimes, may have contributed significantly to the contemporary sediment assemblages now preserved beneath modern flood-plain deposits. The possibility of diverse process mechanisms must be recognised in the reconstruction of palaeo-flow regimes from the sedimentary record. The paper of Maizels (this volume, Chapter 14), discussing sedimentary sequences that may result from both Newtonian and non-Newtonian flow processes in an environment with no *modern-day* analogue in the UK, repays study in this respect.

However, despite initiatives of this kind, evidence for large floods in the Quaternary sequence of sedimentary deposits remains speculative and equivocable. What is diagnostic of a flood deposit? How can flood deposits be expected to reflect the nature of the catchment that produced them? There is clearly tremendous potential to integrate the investigations conducted at scales appreciated by the sedimentologist with scales familiar to the geomorphologist.

Fluvial geomorphology will benefit by the fusion of ideas within a basin-wide framework in which the differing responses of headwater and piedmont reaches are integrated in a conceptualisation of system dynamics (Richards, 1986). In the papers that follow, Newson stresses the need to expect complexity in fluvial systems, and Lewin argues forcibly that it is time to critically re-examine many of the precepts that formed the basis of fluvial geomorphology in the 1950s and 1960s. Nested hierarchies of process–response models need constructing, that combine classical uniformitarian concepts with the new ideas of intermittency of process and response, and link hydrological, sedimentological and geomorphological processes and scales.

Recently, the geological sciences have seen a revolution of thought in respect of catastrophic events and the interpretation of stratigraphic sequences (Ager, 1981). Similarly, in some geomorphological settings, floods may play a vital role in instigating system change in a way that may not be immediately evident in a relatively impoverished terrestrial, as opposed to extensive oceanic, sedimentary record. Ideas have been formulated here and elsewhere (e.g. Huggett, 1988) as to how modelling strategies should proceed by linking theory with field examples in a way that combines the expertise from different disciplines. It is on this basis that we can confidently look forward to a vital fluvial geomorphology in the years leading to the next millenium.

REFERENCES

Ager, D. V. (1981). *The Nature of the Stratigraphic Record* (2nd edn), Macmillan, 122p.
Anderson, M. G., and Burt, T. P. (1986). *Hydrological Forecasting*, Wiley, Chichester.
Anderson, M. G. and Calver, A. (1977). On the persistence of landscape features formed by a large flood, *Trans. Instn. Brit. Geog., NS.*, **2**, 243–54.
Baker, V. R. and Costa, J. E. (1987). Flood power, in L. Mayer and D. Nash (eds), *Catastrophic Flooding*, Binghamton Symposia in Geomorphology Series No. 18, 1–24.
Bathurst, J. C. (1986). Physically-based distributed modelling of an upland catchment using the Système Hydrologique Européen, *J. Hydrology*, **87**, 79–102.
Benson, R. H. (1984). Perfection, continuity and common sense in historical geology, in W. A. Berggren and J. A. Van Couvering (eds), *Catastrophes and Earth History: The New Uniformitarianism*, Princeton University Press, Princeton, New Jersey, 35–75.
Beven, K. J. (1981). The effect of ordering on the geomorphic effectiveness of hydrological events, *IASH Publication 132*, 510–26.
Beven, K. J. (1987a). Towards a new paradigm in hydrology, *IASH Publication No. 164*, 393–403.
Beven, K. J. (1987b). Towards the use of catchment geomorphology in flood frequency predictions, *Earth Surf. Process. Landf.*, **12**, 69–82.
Beven, K. J., Calver, A., and Morris, E. M. (1987). The Institute of Hydrology Distributed Model, *Institute of Hydrology Report 97*, Wallingford, Oxon.
Borah, D. K., Alonso, C. V., and Prasad, S. N. (1982). Routing graded sediments in streams: formulations, *J. Hydraul. Div., ASCE*, **108**, 1486–1503.
Calver, A. (1988). Calibration, sensitivity analysis and validation of a physically-based rainfall-runoff model, *J. Hydrology*, in press.

Carling, P. A. (1987). A terminal debris-flow lobe in the northern Pennines, United Kingdom, *Trans. Royal Soc. Edinburgh, Earth Sciences*, **78**, 169–76.

Cordova, J. R. and Rodriguez-Iturbe, I. (1983). Geomorphologic estimation of extreme flow probabilities, *J. Hydrology*, **65**, 159–73.

Diaz-Granados, M. A., Valdes, J. B., and Bras, R. L. (1984). A physically-based flood frequency distribution, *Water Resources Research*, **20**, 995–1002.

Dury, A. (1973). Magnitude frequency analysis and channel morphometry, in M. Morisawa (ed.), *Fluvial Geomorphology*, Binghamton Symposia in Geomorphology Series, 91–121.

Hebson, C. and Wood, E. F. (1982). A derived flood frequency distribution, *Water Resources Research*, **18**, 1509–18.

Huggett, R. J. (1988). Dissipative systems. Implications for Geomorphology, *Earth Surf. Process. Landf.*, **13**, 45–50.

Hutchinson, W. (1794). *The History of the County of Cumberland and Places Adjacent*, Jollie, Carlisle.

Lyn, D. A. (1987). Unsteady sediment transport modelling, *J. Hydraul. Eng. ASCE*, **113**, 1–15.

Moughamian, M. S., McLaughlin, D. B., and Bras, R. L. (1987). Estimation of flood frequency: an evaluation of two derived distribution procedures, *Water Resources Research*, **23**, 1309–19.

Newson, M. D. (1980). The geomorphological effectiveness of floods, a contribution stimulated by two recent events in mid-Wales, *Earth Surface Processes*, **5**, 1–16.

Ponce, V. M., Garcia, J. L., and Simons, D. B. (1979). Modelling alluvial channel bed transients, *J. Hydraul. Div. ASCE*, **105**, 245–56.

Prigogine, I. and Stengers, I. (1984). *Order out of Chaos*, Heinemann.

Richards, K. S. (1986). Fluvial Geomorphology, *Progress in Physical Geography*, **10**, 401–20.

Smith, G. (1754). Dreadful storm in Cumberland, *Gentleman's Magazine*, **24**, 464–7.

Stephens, M. A., Simons, D. B., and Richardson, E. V. (1975). Nonequilibrium River Form, *J. Hydraul. Div. ASCE*, **101**, 243–54.

Storm, B., Jorgensen, G. H., and Styczen, M. (1987). Simulation of water flow and soil erosion processes with a distributed physically-based modelling system, *IASH Publication No. 167*, 595–608.

Wells, S. G., and Harvey, A. M. (1987). Sedimentological and geomorphic variations in storm-generated alluvial fans, Howgill Fells, northwest England, *Bull. Geol. Soc. Amer.*, **98**, 182–98.

Williams, G. P. (1978). Bankfull discharge of rivers, *Water Resources Research*, **14**, 1141–54.

Wolman, M. G., and Gerson, R. (1978). Relative scales of time and effectiveness of climate in watershed geomorphology, *Earth Surface Processes*, **3**, 189–208.

Wolman, M. G., and Leopold, L. B. (1957). River flood plains: some observations on their formation, *USGS Prof. Paper*, **282c**, 87–107.

Wolman, M. G. and Miller, J. P. (1960). Magnitude and frequency of forces in geomorphic processes, *J. Geology*, **68**, 54–74.

2 Storm Runoff Generation in Small Catchments in Relation to the Flood Response of Large Basins

T. P. BURT
School of Geography, University of Oxford, Oxford

INTRODUCTION

Until very recently, engineering hydrologists took the view that the headwaters of a drainage basin were nothing more than the upstream source areas for runoff which duly generated floods at downstream locations. Since their concern was with flood forecasting at the lowland site, the physical characteristics of the headwater region and the hillslope processes responsible for producing the runoff could be safely ignored. Since most flood modelling involved a 'lumped' or aspatial approach, at best the catchment characteristics were described by simple summarial parameters such as area, main stream length, mean channel slope, and so on. In many cases, of course, lumped models provide perfectly acceptable flood forecasts despite their reliance on such simple measures; a good example is the synthetic unit hydrograph model produced by the UK Flood Studies Report (National Environment Research Council, 1975).

Sherman produced his unit hydrograph model in 1932. This was a major advance in rainfall-runoff modelling and the technique was widely adopted. At much the same time, Horton (1933) produced his classical model of hillslope hydrology. Horton's simplistic approach assumed that the sole source of storm runoff was the excess water which was unable to infiltrate into the soil. Water which infiltrated formed the sole source of baseflow. As Chorley (1978) notes, it was 'a particularly happy circumstance' that Horton's model fitted so well with Sherman's highly influential unit hydrograph model. Horton's ideas provided a physical basis for Sherman's lumped, empirical approach. Together, their ideas dominated catchment hydrology for several decades.

As described below, the work of John Hewlett ushered in a new phase of research on hillslope hydrology. Despite the major advances in our understanding of physical hydrology which resulted, flood forecasters continued for some years to prefer lumped models of the drainage basin, often based on a reservoir concept to allow some physical reality to be incorporated into the model structure. However, over the last decade, physically based distributed models have been developed (Beven, 1985). These have the capability of forecasting the spatial pattern of hydrological conditions within a catchment as well as the discharge outflow and bulk storage. It is possible to identify at least five major areas of application which distributed models may be used to forecast: the effects of land-use change; the effects of spatially variable inputs, catchment characteristics, and outputs; the movement of sediments and solutes through the basin; extreme events; and, the response of ungauged basins. Thus, the use of distributed models requires attention to process and to location. In meeting this requirement, knowledge of hillslope hydrology becomes crucial to any attempt to predict flood response at the basin scale.

STORM RUNOFF MECHANISMS

Controls of storm runoff generation

The Variable Source Area was first defined by John Hewlett (1961). Since then, his concept has come to dominate hillslope hydrology. Subsurface flow is now viewed as *the* major runoff-generating mechanism, both because of its influence on saturation-excess overland flow (Dunne and Black, 1970), and as an important contributor to stormflow in its own right (Anderson and Burt, 1978). The idea of subsurface stormflow is not new (Hursh and Brater, 1941), but the dominance of runoff theories based on the occurrence of infiltration-excess overland flow (Horton, 1933, 1945) meant that research into subsurface flow mechanisms was neglected. However, despite modifications such as the Partial Area concept (Betson, 1964), it became apparent that Horton's model was inappropriate in many locations. In areas of permeable soils, where there is a decrease in hydraulic conductivity with depth, subsurface runoff within the soil can account for much if not all of the stormflow leaving a catchment. When the soil profile becomes completely saturated, saturation-excess overland flow will also occur. Both processes may occur at rainfall intensities well below those required to produce infiltration-excess overland flow (Zaslavsky, 1970; Burt, 1986). In addition, both processes are generated from source areas which may be variable in extent and different in location from the source areas for 'Hortonian' overland flow.

Despite much research, the production of storm runoff in highly responsive headwater catchments is still poorly understood. For example, it is only within the last few years that the importance of macropore flow has been finally recognised (Beven and Germann, 1982). Another example of our lack of understanding of storm runoff mechanisms relates to the fact that the chemical composition of storm runoff is often very different from that of the precipitation. As noted below, a number of researchers have identified the relative contributions of 'old' and 'new' water to the storm hydrograph, but the precise manner in which translatory flow occurs (i.e. lateral displacement of soil water into the stream) has still not been clearly defined. Also, it should be remembered that infiltration-excess overland flow is not necessarily as rare as proponents of subsurface stormflow have argued. There are situations, even on apparently permeable soils, where this process can occur—probably because the soil surface has become compacted by overgrazing, by heavy machinery, or by rainbeat. When it does, it is often accompanied by high rates of topsoil erosion (Burt, 1987). Figure 2.1 describes the various mechanisms involved in the generation of storm runoff. These will be briefly reviewed below, though it should be remembered that the relative importance of some processes is still the subject of debate, and that the detailed mechanics of some processes have yet to be fully specified.

As Dunne (1978, Figure 7.40) has noted, the dominant controls of storm runoff generation are climate and soils, with topography as an important secondary control at the subcatchment scale. Kirkby (1978) identified the crucial role of hydraulic conductivity in relation to rainfall intensity (Figure 2.2). He recognised domains dominated either by infiltration-excess overland flow or by the combination of subsurface stormflow and saturation-excess overland flow. Of course, in some situations, certain soils will straddle the boundary between these two domains, especially where surface deterioration lowers the infiltration capacity of what would otherwise be a very permeable soil (Burt, 1987). Burt and Butcher (1986) have produced computer simulations to illustrate these domains (Figure 2.3).

Freeze (1986) has also used a simulation model to consider the interaction of climate and soil with respect to mass movement processes and the stable angle of

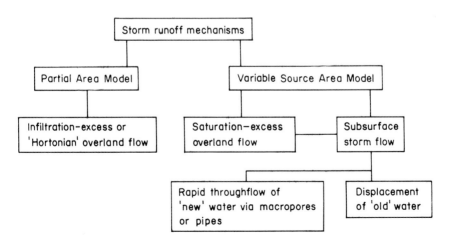

Figure 2.1 Storm runoff mechanisms.

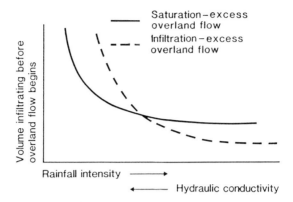

Figure 2.2 The relationship between the volume of rainfall which infiltrates before overland flow begins and rainfall intensity. Soils with low hydraulic conductivity will be dominated by infiltration-excess overland flow; those with high hydraulic conductivities by saturation-excess overland flow (After Kirkby, 1978).

hillslopes. His results are relevant both to runoff and sediment production (both by mass movement and by particulate erosion). Freeze shows that hydraulic conductivity is a parameter of some importance in that it controls both the occurrence of infiltration-excess overland flow, and the development and extent of saturation within the soil profile. As Table 2.1 shows, in arid areas there is surface runoff only where the soil or bare rock surface is impermeable or where storage is very limited; the Partial Area model of runoff generation appears most appropriate (Yair and Lavee, 1985). In humid areas, many studies have shown

Table 2.1 Runoff production in relation to climate and soil.

Climate	Arid	Humid
Permeable	No runoff except where storage is very limited, or where colluvial infill in gullies lowers infiltration capacity (Yair and Lavee, 1976, 1986)	Saturation-excess overland flow is most important, except where deep permeable soils directly abut the stream, in which case sub-surface stormflow is dominant (Dunne, 1978; Burt, 1986).
Impermeable	Saturation of surface horizons only. Rilling indicates infiltration-excess overland flow (Schumm, 1964); shallow mass movements denote surficial soil saturation (Freeze, 1986). Bare rock surfaces can be major Partial Areas (Yair and Lavee, 1986).	Infiltration-excess overland flow is most likely, although in low intensity rainfall saturation-excess overland flow can also occur (Burt and Gardiner, 1984). Subsurface stormflow is limited to pipes and macropores.

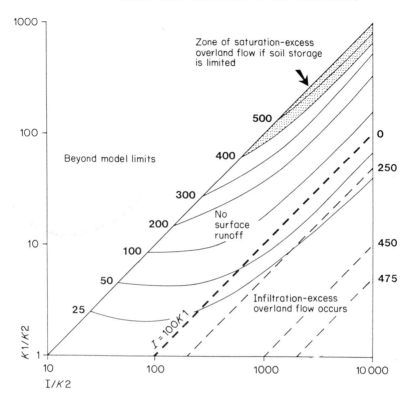

Figure 2.3 Hillslope runoff production in relation to the ratio $K1/K2$ (where $K1$ and $K2$ are the hydraulic conductivities of the upper and lower soil horizons) and the ratio of rainfall intensity (I) to $K2$. Solid lines refer to total volumes of subsurface runoff; dashed lines refer to total volumes of infiltration-excess overland flow. The results were obtained using a simple hillslope runoff model (From Burt and Butcher, 1986, by kind permission of the *Journal of Geography in Higher Education*).

that the frequency and magnitude of storm runoff is controlled by the extent of saturated areas—as Hewlett (1961) showed, these are 'variable' in time and space. Such areas can respond rapidly even to low intensity rainfall. Burt and Butcher (1985a) have extended the Variable Source Area model to include the generation of subsurface stormflow. They show that the size of the saturated wedge contiguous to the stream controls the magnitude of the throughflow response. Thus, unlike arid areas, in humid areas the spatial distribution of soil moisture is also an important control of storm runoff generation, and the spatial non-uniformity of runoff production is related both to infiltration capacity and to soil moisture distributions.

As Hewlett has shown, the location of Variable Source Areas is mainly determined by topography; as explained below, their existence depends on the downslope movement of soil moisture, with hollows being a particular focus for the development of saturation. In this respect, the role of topography, whilst secondary to climate and soil, is crucial to an understanding of the spatial pattern of storm runoff generation in humid drainage basins.

Infiltration-excess overland flow

The Partial Area model proposed by Betson (1964) remains the best guide to the location of source areas for infiltration-excess overland flow. One difficulty though, is to predict which partial areas, not contiguous to the channel, can introduce runoff and sediment to the stream; Jones (1979) has termed such distant sources 'disjunct contributing areas'. Another difficulty is that, despite much effort, observations of overland flow remain rare. For example, Emmett (1978) notes that during a decade of research in semi-arid New Mexico, overland flow was not observed once, although it was clear from measurements that surface erosion on the unrilled slopes accounted for 98% of the total sediment yield. This lack of field evidence has made it more difficult to produce a general description of the process. Where observations have been made, as in various badlands, a further complication is that runoff may occur as shallow subsurface flow through macropores, rather than 'classic' overland flow (Bryan et al., 1978).

As Emmett (1978) points out, overland flow is both unsteady and spatially varied since it is supplied by rain and depleted by infiltration, neither of which is necessarily constant with respect to time and location. This is one reason why theoretical descriptions of the infiltration process have been difficult to aggregate to the hillslope scale. It is crucial to describe the generation of infiltration-excess overland flow for two reasons: it can produce the highest peak runoff with the shortest response times (Dunne, 1978) and is therefore vital to an understanding of storm hydrographs. Also, overland flow is capable of eroding the slope in a variety of ways—inter-rill and rill erosion, and gullying (Meyer, 1986)—and thus may introduce significant sediment load to the stream. Most considerations start from the quantitative approach of Horton (1945) but Dunne and Aubry (1986) have shown that modifications are needed to his original framework to take account of rainsplash, a point also made by De Ploey (1984). Dunne and Aubry (1986) show that the initiation and maintenance of rills depends upon the balance between sediment transport by sheetwash, which tends to cause channel incision, and rainsplash, which tends to infill channels, thereby smoothing the surface. Rills form on a sufficiently steep slope when the sheetflow becomes unstable and starts to incise, or where a sufficient length of slope exists to provide an unstable flow. Dunne and Aubry stress that a lack of rills does not imply a lack of erosion, as Horton (1945) argued; the maintenance of rills depends on the balance of splash and overland flow. However, as Meyer (1986) shows, rilling adds to the inter-rill erosion and so can increase sediment inputs to stream channels (Figure 2.4).

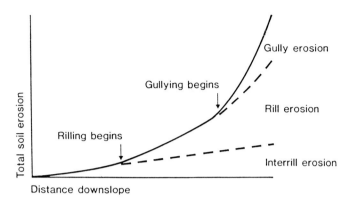

Figure 2.4 The location of interrill, rill and gully erosion (After Meyer, 1986).

Meyer (1986) has reviewed the physical characteristics of eroded sediment and the factors which affect soil erosion rates. The best-known equation for quantifying erosion rates is the Universal Soil Loss Equation (USLE) which estimates annual soil loss caused by rill and inter-rill erosion (Wischmeier and Smith, 1965, 1978). Modifications to USLE take gully erosion into account. The factors incorporated into USLE summarise the main controls of soil erosion: erosiveness of rainfall; erodibility of soils; slope length and steepness; cropping and management techniques; and associated conservation practices. For the present, empirical approaches like USLE remain the best available method for predicting sediment yields to streams, although as Walling (1983) points out, it is increasingly necessary to disaggregate the 'black box' of sediment delivery. Dunne and Aubry (1986) argue that deterministic models of sediment transport by overland flow are not yet accurate predictors, and that they must be improved on the basis of theoretical and experimental study. It remains to be seen how best the approaches of soil erosion engineers and hydrologists can be combined into an acceptable, distributed runoff-erosion model at the hillslope or subcatchment scale.

Subsurface stormflow

As Figure 2.1 shows, subsurface stormflow may be generated by two mechanisms: by non-Darcian flow through macropores or pipes (which are somewhat arbitrarily distinguished on the basis of size); and by Darcian flow through the soil matrix. Hillslope hydrologists have until recently emphasised matric flow, although there has been a continuing interest in pipeflow (Jones, 1981). Field evidence relating to macropores (e.g. Whipkey, 1965) was somewhat ignored until recently, when there has been renewed interest in preferential flow through structured soils. Subsurface flow may provide stormflow in the following ways:

(i) Pipeflow

A number of studies have described the occurrence and hydrological function of pipes (see reviews by Gilman and Newson, 1980, and by Jones, 1981). As Jones (1979) noted, the rapidity of pipeflow is so great that contributing areas distant to the stream can generate stormflow. Bonell *et al.* (1984) showed, for a forested clay soil in Luxembourg, that pipeflow in the saturated upper soil horizon was so rapid that it could not be distinguished from saturation overland flow. The effect of underdrainage on the hydrological response is discussed by Robinson and Beven (1983) and by Reid and Parkinson (1984a, 1984b).

(ii) Macropore flow

Following the publication of several catalytic articles (e.g. Mosley, 1979; Beven and Germann, 1981, 1982), a number of recent studies have emphasised the hydrological effects of macropore flow. Reid and Parkinson (1984a; 1984b) and Kneale (1986) have shown the importance of cracking to the soil water regime of a clay soil. Kneale and White (1984) studied infiltration into 9 cm cores of dry, cracked clay–loam topsoil. They showed, for rainfall intensities above the pedal infiltration capacity of 2.2 mm h^{-1}, that bypassing flow occurred down the cracks. Although the absolute magnitude of absorption of water by the peds increased with higher intensity, so too did the ratio of bypass flow to absorption, to reach an output:input ratio of 55% at an input intensity of 21.9 mm h^{-1}. For summers in the Oxford (UK) region, where the soils were collected, this implies that between 10% and 20% of rainfall would bypass the root zone. At the same site, Kneale (1986) estimated the hydraulic conductivity of soil peds to be only 1.4 mm h^{-1}; by contrast, the bulk hydraulic conductivity of the topsoil is 1800 mm h^{-1}.

Germann (1986) and Coles and Trudgill (1985) have identified important thresholds controlling macropore flow (Figure 2.5). Both studies emphasise the importance of the pedal infiltration capacity: at lower rainfall intensities no by-passing flow is generated *except*, as Coles and Trudgill demonstrate, when the soil is at field capacity, when no more water can be stored in the peds and bypassing must occur. Both studies also identify a further threshold of antecedent soil moisture: if the soil is too dry then any flow in the macropores is rapidly absorbed into the matrix and no by-passing flow is generated. Germann showed that the upper metre of soil must have a moisture content of at least 30% by volume, a figure very close to the threshold found by Coles and Trudgill. Such results have clear implications for storm runoff production. Robinson and Beven (1983) showed, for a swelling clay soil, that flow along cracks produced

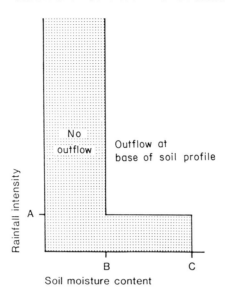

Figure 2.5 Thresholds controlling macropore flow. Point A indicates the infiltration capacity of soil peds; points B and C indicate soil moisture levels below which no outflow occurs, depending on rainfall intensity (After Coles and Trudgill, 1985, and Germann, 1986).

higher peak flows compared to an uncracked soil, in summer when the soil was dry.

One question which remains unsolved is the connectivity of macropores in the *downslope* direction. Studies such as Whipkey (1965), Pilgrim *et al.* (1979) and Imeson *et al.* (1984) demonstrate that macropores can generate rapid downslope runoff as well as aiding infiltration. In other cases, their downslope connectivity is not proven. Pearce *et al.* (1986) studied runoff generation at a site in Tawhai State Forest, New Zealand. The fact that only 3% of storm runoff could be considered 'new' cast doubt on the findings of Mosley (1979) who had emphasised macropore flow at the same location. Further research using tracers, such as dyes (Trudgill, 1986) or natural isotopes (Pearce *et al.*, 1986), is required to resolve this issue. Macropore flow may well occur laterally at Tawhai, but clearly some translatory flow mechanism is involved too.

(iii) Flow through the soil matrix

Classical infiltration models have assumed a semi-infinite soil where storage effects are unimportant (Horton, 1933, 1945; Philip, 1957); overland flow occurs only when rainfall intensity exceeds infiltration capacity. In layered soils, where hydraulic conductivity declines with depth, storage is limited: overland flow can occur even though rainfall intensity is well below infiltration capacity, once the

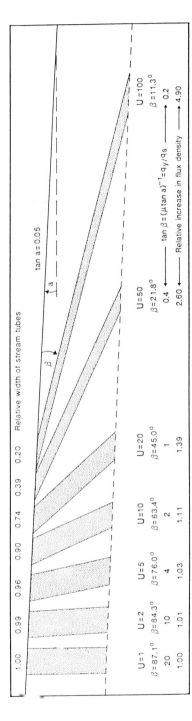

Figure 2.6 Flow direction as controlled by soil anisotropy. U give the ratio of horizontal to verticle hydraulic conductivity; also shown is the resultant flow direction. The relative increase in flux density is shown by the width of the stream tube (From Zaslavsky, 1970).

upper soil layer becomes saturated. Zaslavsky (1970), Zaslavsky and Sinai (1981) and Burt (1986) have considered the relation between soil layering and the generation of lateral subsurface flow. As the ratio between the hydraulic conductivity of the upper and lower soil layers increases, so the flow direction becomes more nearly parallel to the slope (Figure 2.6). In itself, the occurrence of lateral subsurface flow is not enough to produce stormflow. It is often assumed that matrix flow is too slow to provide this.

However, rapid subsurface flow can occur through a permeable upper horizon if the soil is saturated, or if the water table is close enough to the soil surface so that the 'capillary fringe' extends to the surface, or close to it (Abdul and Gillham, 1984; Gillham and Abdul, 1986). They define the capillary fringe as the unsaturated zone, above the water table, but below the point where soil drainage occurs. If the fringe extends to the surface then only a small amount of water is needed to produce a significant rise in the water table, much greater than might be supposed given the specific yield of the soil and the magnitude of the rainfall input. A higher water table increases the hydraulic gradient and causes lateral flow, in contrast to the previously vertical drainage of the unsaturated zone. Only when the capillary fringe is well below the surface will 'normal' infiltration occur with a consequent delay in the subsurface response. If the saturated hydraulic conductivity of the soil is high, then large amounts of subsurface stormflow will be rapidly generated as a result of a rise in the water table. This mechanism is sometimes described as 'groundwater ridging' (Sklash and Farvolden, 1979). Most of this subsurface flow will be 'old' water displaced from the base of the slope. This process is identical to the translatory flow mechanism originally described by Hewlett (1961) and by Hewlett and Hibbert (1967), and to the 'piston flow' effect described by Anderson and Burt (1982).

In addition to this immediate effect, subsurface stormflow may also occur in the form of a delayed hydrograph peaking several days after the rainfall input (Burt and Butcher, 1985b). Such hydrographs are usually the major volumetric response and in some cases may provide the peak discharge too. Lateral subsurface flow is clearly crucial to the generation of such delayed hydrographs. The same topographic factors which control the location of saturation-excess overland flow are important here (see below). Such runoff is strongly seasonal, being largely confined to the winter half of the year when soil moisture deficits have been nullified (Beven, 1982; Burt, 1987). Figure 2.7 shows the surface and subsurface flow components for a double-peaked hydrograph. Subsurface flow responds rapidly at the time of the rainfall and contributes most of the first discharge peak. The second peak is entirely subsurface flow. Thus, subsurface stormflow can be produced in significant quantities and in some basins will dominate the total stormflow response (Dunne, 1978; Burt, 1986).

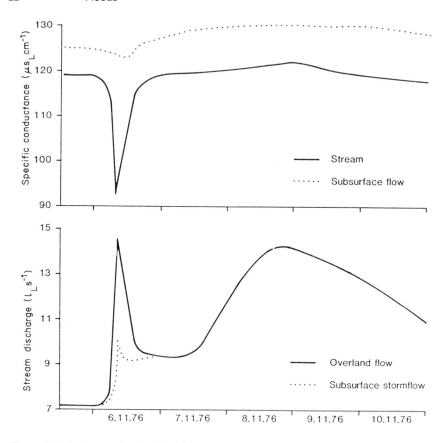

Figure 2.7 Surface and subsurface flow components at the Bicknoller Combe catchment.

Saturation-excess overland flow

It is impossible to divorce the generation of subsurface stormflow from the production of saturation-excess overland flow. The Variable Source Area model, in which saturated zones expand upslope during the storm, is based upon the assumption of lateral translatory flow through the soil (Hewlett and Hibbert, 1967). The source areas for the two types of stormflow are thus essentially identical. The spatial distribution of soil moisture is mainly controlled by topography. Three zones may be identified where maximum soil moisture levels will be reached (Kirkby and Chorley, 1967; Burt, 1986). Perhaps the most important are hillslope hollows where convergence of flow lines favours the accumulation of soil water (Anderson and Burt, 1978). As well as being a major source of subsurface flow, they will also be the location where surface saturation is most likely to occur (Dunne, 1978). At the slope base levels of soil saturation is are highest for a given slope profile, both because runoff increases downslope and because many slopes are concave at their base, which also encourages the accumulation of soil water. In areas of reduced storage, the transmissivity of the soil profile is reduced and moisture levels can be expected to rise.

Kirkby (1978), O'Loughlin (1981, 1986), Burt and Butcher (1985a) and Thorne *et al.* (1987) have generalised about the distribution of soil moisture using a variety of topographic indices; in all cases, upslope drainage area is the crucial control. The extent of the saturated area depends on soil wetness and so source areas vary seasonally and during storms (Dunne, 1978). Saturation-excess overland flow will be a mixture of return flow (translatory flow) and direct runoff (rain unable to infiltrate into the saturated ground). Where surface saturation occurs to any great extent, saturation-excess overland flow will dominate the stormflow response with higher peak discharge and lower lag times than are characteristic for subsurface stormflow (Dunne, 1978). Even so, with permeable soils the zone of saturation will be very restricted so that runoff:rainfall ratios of less than 5% are common. However, in humid areas, for soils of low hydraulic conductivity, saturated areas will be much larger and a much higher percentage of rainfall will be translated into stormflow (Burt and Gardiner, 1984).

Even where the ratio of storm runoff to storm rainfall is low, dramatic floods can still be produced. This has been shown by Hewlett *et al.* (1977) for the Appalachian region of the United States. Here, most areas have a mean ratio below 0.25; even so, flooding is a major hazard in the region. Small increases in this ratio caused by changes to the catchment such as forest clearance may therefore have a marked effect. Furthermore, in such catchments the contributing areas may expand greatly at certain times, particularly when high rainfall has preceded a given storm or during very large rainfall events. The result is that extreme flood events in such catchments are likely to be associated with extensive development of surface saturation. Such variability in the size of source areas has been described schematically many times, though examples such as those described in Dunne (1978) are surprisingly rare. O'Loughlin (1986), in generalising about the spatial distribution of soil moisture and its relationship to flood runoff generation, has shown that source areas are essentially 'fixed' when a basin is dry, since additions to soil moisture storage have little effect on the extent of surface saturation. However, for wet antecedent conditions, a small amount of infiltration can greatly extend the runoff source area, which is now quite clearly 'variable'. Clearly it becomes crucial to incorporate this element of non-linearity into flood forecasting models. Through the use of models which realistically consider the runoff processes operating and the locations where flood runoff is produced, these elements of spatial and temporal variability may be most easily included.

SEDIMENT AND SOLUTE PRODUCTION

Storm runoff mechanisms determine the delivery of sediment and solutes to the stream. Walling (1983) has noted that most studies of sediment delivery systems are 'lumped' with respect to time and space. In order to take a 'distributed' approach, reference to hillslope hydrology is vital.

Sediment delivery by overland flow

Both the Variable Source Area and Partial Area models need to be invoked if the spatial pattern of sediment delivery is to be fully described. For saturation-excess overland flow, surface runoff is produced from only a small proportion of the basin. Since the size of the contributing area varies with antecedent soil moisture conditions, the amount of sediment delivered to the stream depends on the characteristics of the source area at any given time. During extreme storms, a greatly expanded contributing area could mobilise sediment in areas which are unconnected to the stream under normal conditions (Walling, 1983). However, there is a lack of field evidence that demonstrates this link between runoff and sediment production; Moore *et al.* (1986) discuss the problem theoretically.

Surface runoff production from partial areas is largely controlled by infiltration capacity. Low infiltration is often associated with bare soils which may occur naturally, as in badlands (Yair *et al.*, 1980; Bryan and Campbell, 1982), or under agriculture (e.g. Boardman, 1983; Colborne and Staines, 1985; Fullen and Harrison-Reed, 1986). Perhaps the main problem here is to identify possible linkages between the source of soil erosion and the stream; recent evidence suggests that distant sources may connect via roads and tracks, a point not previously emphasised (Burt, 1987). A further research requirement is to examine soil erosion in relation to rainfall intensity with durations of only a few minutes. Often, analysis is confined to hourly intensities, but significant erosion may be achieved in short bursts of very intense rain. Colborne and Staines (1985) discuss possible changes in agricultural practice which might reduce soil losses: this is important, not only in its own right, but because sediment-bound pollutants such as phosphate and pesticides (Willis and McDowell, 1982) are an increasing cause for concern. The relationship between runoff processes and the initiation of natural drainage networks, and the implications of this for sediment delivery, have been reviewed by Dunne (1980) and Jones (1987).

Sediment delivery by mass movement

As Freeze (1986) has shown, there is an intimate relation between soil moisture levels and the occurrence of mass movements. This is particularly important with respect to convergent flow processes in hillslope hollows. Tsukamoto *et al.* (1982) noted that more than 80% of debris slides observed in the Japanese mountains occurred on convergent slopes. Indeed, they have called such slopes zero-order basins, being the most important from a hydrological and geomorphological point of view. Dietrich and Dunne (1978) and Dietrich et al. (1986) discuss this aspect further. As Newson (1980) has shown, the delivery of sediment to the stream channel by mass movement can markedly affect the geomorphological effectiveness of floods, both at the time, and for future floods.

Solute delivery

Current concerns about the effects of acid precipitation, including the episodic acidification of some rivers during flood events, and about the leaching of fertilisers has heightened interest in the links between storm runoff mechanisms and solute leaching. There is often the call for a better formulation of the catchment hydrology (e.g. Pearce et al., 1986; Reynolds et al., 1986; Lynch et al., 1986). A number of studies have used chemical or isotopic methods to divide storm runoff into 'old' and 'new' water (e.g. Pilgrim et al., 1979; Sklash and Farvolden, 1979; Anderson and Burt, 1982; Kennedy et al., 1986; Pearce et al., 1986). A surprisingly small percentage of storm runoff may prove to be 'new' water. However, the translatory flow has not been closely documented: empirical evidence exists which correlates soil moisture movement with the hydrographs and chemographs of throughflow and streamflow (Burt, 1979, 1984; Burt et al., 1983) but direct evidence based on soil moisture samples from suction lysimeters and piezometers has proved difficult to acquire. Figure 2.7 shows that delayed subsurface stormflow can be accompanied by an increase in solute concentration. In this example the effect is small; in other instances much larger increases in concentration have been identified which, in the case of nitrate, have exceeded drinking water standards (Burt et al., 1983).

The spatial pattern of stream solutes depends, at the basin scale, on climate, lithology and land use (Walling and Webb, 1975). At the subcatchment scale, Burt and Arkell (1986; 1987) have shown that land use and topography are important controls. Table 2.2 shows flow contributions for part of the Slapton Wood catchment in Devon, England. Stream reaches are classified into hillslope hollows and valleyside slopes. The latter are important *non-point* sources of discharge, being fed by steep slopes with a narrow interfluve. Hollows, despite the importance of convergent flow, include large areas of interfluve which contribute little runoff. Thus the valleyside slopes have a larger unit area runoff contribution (dQ/A). However, hollows are clearly important *point* sources of discharge contributing 47% in only 7.5% of the channel length (dQ/L). Examining discharge increments by slope area *and* channel distance shows that hollows are the dominant contributing areas (dQ/AL). Since the hollows drain most of the intensively farmed land, their unit area contribution of nitrate is more similar to the valleyside slopes (dN/A). This emphasises the role of hollows as point sources of nitrate pollution, although the non-point sources are

Table 2.2 Flow contributions in the Slapton Wood catchment on 26.2.84.

Reach type	dQ	dQ/A	dQ/L	dQ/AL	dN/A	dN/L	dN/AL
Hollow	6.07	2.67	11.60	5.15	3.16	1.37	6.03
Valleyside	6.70	4.64	1.59	1.10	3.76	0.13	0.89
Hollow	4.83	2.51	15.3	7.49	2.24	1.38	7.12
Valleyside	6.50	7.29	2.06	2.31	4.45	0.13	1.41

Abbreviations: dQ = change in stream discharge over stream reach (1 s–1)
 dN = change in nitrate load over stream reach (mg s–1)
 A = hillslope area draining to stream reach (m2)
 L = length of stream reach (m)

Units: dQ/A = 1 s–1 m–2 × 10–5
 dQ/L = 1 s–1 m–1 × 10–2
 dQ/AL = 1 s–1 m–2 × 10–7
 dN/A = mg s–1 m–2 × 10–4
 dN/L = mg s–1 m–1
 dn/AL = mg s–1 m–1 m–2 × 10–7

important too. Comparative data are lacking for flood events, but it seems likely that the relative importance of hollows is further increased when delayed hydrographs occur. Controlled experimentation is planned to discriminate between the effects of land use and topography: in this example, the two interact, with hollows draining mainly arable land and valleyside slopes being woodland or grass. At the same time, this work aims to establish the exact mechanisms involved in solute delivery to the stream.

FLOOD RUNOFF PRODUCTION IN LARGE BASINS

Despite great interest in hillslope hydrology over the last two decades, the knowledge gained has made less impact on the broader field of flood forecasting than one might have hoped. The development of distributed flood forecasting models has been slow, partly because model descriptions need to be improved, and also because of the cost of obtaining field data for model calibration. However, as Beven (1985) notes, distributed models will be increasingly used in the future, the incentive to do so being that they are 'better' models in the sense of having a more rigorous theoretical basis, and because they have the capability to forecast the spatial pattern of hydrological conditions within the basin as well as simple outflows and storages. At present, however, such models remain prototypes so that the opportunity to use them for simulation experiments to study, for example, the effects of changes in one part of the basin on the flood hydrograph of the whole basin, has been rare.

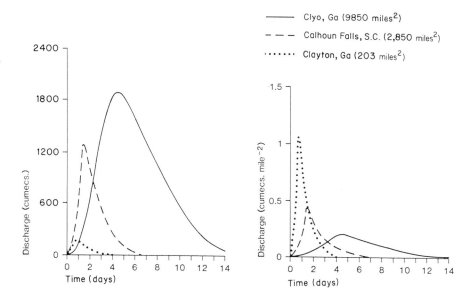

Figure 2.8 Flood hydrographs for three ganging stations on the Savannah River, USA. Plotted simply as rate (left), the record shows how the peak discharges upstream are 'lost' in the volumes of water which created the flood at Clyo (After Hewlett, 1982).

There is also surprisingly little empirical evidence that can be used to relate small catchments, where hillslope runoff generation is the dominant hydrological mechanism, to the flood response of large basins, where channel storage and routeing must also be considered. Most hydrologists are familiar with hydrographs such as those shown on Figure 2.8: whilst the absolute size of the hydrograph predictably increases downstream, the unit-area peak discharge decreases downstream. Consider Hewlett's conclusions on this matter:

> On the main stem of a river, peak discharges do the flood damage. However, *it is not the peak discharge in the headwaters that produces the downstream flood, but rather the volume of stormflow released by the headwater areas.* That hydrologic fact cannot be overemphasised. As many first-order basins contribute stormflow, their respective peak discharges are staggered in time so that tributary *peak rates* are not additive downstream, whereas tributary *volumes* are additive downstream. Once the water is in the channel system, land use above obviously has no further influence on the part that particular water plays in the flood stage below. (Hewlett, 1982, p. 163.)

In large basins it is clear that the effects of travel time and channel storage do prevent the establishment of a clear relationship between runoff from small

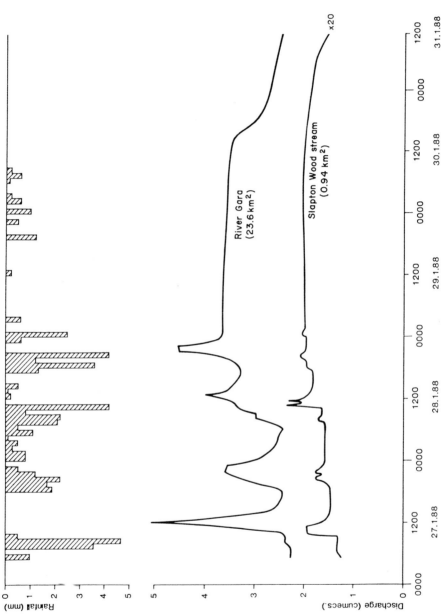

Figure 2.9 Flood hydrographs for the River Gara, Devon, England, and its first-order tributary, the Slapton Wood stream. Note how the delayed throughflow hydrograph of the Slapton Wood stream is reflected in the response of the larger basin.

basins and the flood response of the entire basin. However, Hewlett's statement does require some qualification. It is important to establish the size of basin within which headwater hydrographs *do* correlate with the total basin hydrograph. If a basin is not too large, then flood peaks will be produced at much the same time from all first-order tributaries, and since channel routing is unimportant, the total basin response will represent additions of runoff from individual source areas. Such a case is shown in Fugure 2.9: the runoff response from a small basin (area: 0.2 km^2) remains recognisable at the large scale (area: 23.6 km^2). By implication, if the response of a given basin is effectively the sum of inputs from individual source areas, then the specific location of the source areas may also prove important in that case.

It follows, since changes in land use can affect the volume of storm discharge produced from small basins, that these changes may prove significant at the larger scale. Hewlett and Helvey (1970) showed that stormflow volumes increased by up to 10% when forest was cleared. By implication, major forest clearance could result in more flooding, though as Hewlett notes, the long-running controversy in the USA as to whether forests prevent flooding is by no means proven. Given such presumptions, few researchers have considered whether the *location* of land use changes can affect the flood hydrograph of the main basin. One exception is Acreman (1985) who studied the effects of afforestation on the flood hydrology of the Upper Ettrick valley in Scotland. In upland Britain, poor peaty soils require drainage prior to planting trees. The normal method used is deep ribbon ploughing at right angles to the contours. Such drainage is known to increase the peak discharge and reduce the time-to-peak of flood hydrographs in small catchments (Robinson, 1986). For the Ettrick above Brockhoperig (area 37.5 km^2), Acreman showed that when the lower basin was ploughed between 1968 and 1971, the flood hydrograph became 'flatter'. The unit hydrograph peak flow decreased from 16.1 to 13.8 cumecs with time-to-peak increasing from 3.4 to 4.2 hours. Improved drainage delivered the storm runoff from that area too quickly for it to contribute to the main peak. By contrast, when a similar area of the upper basin was ploughed between 1971 and 1973, the flood response of the whole basin became quicker with higher peak discharges. The unit hydrograph peak flow increased to 22.2 cumecs with time-to-peak reduced to 3.2 hours. Acreman concludes that

> although the small scale effects of afforestation and pre-planting ditching appears to be quite clear cut, the cumulative regional effects of sub-basin land use changes produce a complex modification of the runoff regime by varying the differential timing of flood peaks from different parts of the basin (Acreman, 1985, p. 98).

Such changes to the flood hydrograph may be significant, at least in basins of similar size to the Ettrick. Prediction of such effects could be achieved using distributed models, though this author is unaware of any such simulation experiments.

As Walling (1983) has noted, detailed understanding of sediment delivery to the basin outlet depends in part upon a knowledge of sediment source areas within the basin; similar comments could equally be made with respect to solute delivery. Not surprisingly, however, even less seems to be known about the links between the quality of storm runoff in a small catchment and the quality of flood waters in the larger basin, though Walling and Webb (1980) have shown how dilution of flood waters on the River Exe depends on the origin of the runoff, low concentration runoff from Exmoor producing the greatest dilution downstream. Again, if the effects of land use changes are considered, the importance of linkage between small and large basins is immediately relevant to a number of current pollution problems. For example, the British government is currently considering how to de-intensify agriculture in order to avoid the production of crop surpluses. What would happen to the quality of flood runoff if part of a catchment were 'set aside' from intensive agriculture and allowed to lie fallow for a number of years? The transport of nitrates, pesticides, organic wastes and topsoil might all be affected by such a policy. In upland catchments in Britain, coniferous plantation means not only a change to the flood runoff response, as discussed above, but may also cause acidification of surface waters. Once again, little research has been carried out from a distributed point of view. Welsh *et al.* (1986) studied the acidification of the River Cree in south-west Scotland. Measurements of water quality and fishery evidence supports the view that coniferous afforestation in this poorly buffered catchment has led to to a decrease in the pH of the river. Short-term acid episodes have been identified during periods of flood runoff, some of which may be associated with high sea-salt concentrations (Langan, 1986), or with melting of snowpacks in which acid deposition has accumulated (Warren, 1986). What is *not* clear is how the acidification might depend on which areas of the catchment are forested, and whether there is a threshold area above which pH changes will rapidly ensue. The addition to a distributed runoff model of a geochemical component capable of modelling the effects of conifer plantation and acidic deposition on stream water acidity might usefully address such questions.

It is also worth considering, where the specific location of catchment changes *is* important, that certain zones may be particularly sensitive in that changes here may cause significant changes to flood runoff and water quality downstream. As noted above, variable source areas are partly responsible for non-linearity in the ratio of storm runoff to storm rainfall; extreme floods may occur when there is extensive development of surface saturation throughout the basin. Other sensitive areas include riparian zones and those (partial) areas contributing infiltration-excess overland flow. It is crucial to recognise these areas and to include them in any distributed flood runoff model.

Thus, we can expect that land use changes at the small scale may, in some instances, modify the overall basin hydrograph. Such local changes may also influence water quality during floods. It is not clear how large the area of change must be in order to produce significant modification to the flood response of a larger basin; nor is it clear how much this modification depends upon the *location* of such changes within the larger basin. Sound management of drainage basins may well require a distributed approach in the future, rather than relying on the traditional lumped view which implies that local changes are insignificant. In this regard, the use of distributed flood runoff models may prove essential.

SUMMARY AND CONCLUSIONS

(1) At the basin scale, the major controls of storm runoff generation are climate and soil, with topography important at the subcatchment level.
(2) Both the Partial Area and Variable Source Area models must be invoked in order to relate stormflow production to sediment and solute delivery; indeed even in one small basin, both models may well be applicable.
(3) If the distributed nature of sediment and solute delivery is to be better understood, several points require further study, including: patterns of soil erosion in relation to the characteristics and extent of variable source areas; linkage of partial areas to the main channel system; and, translatory flow mechanisms and the contribution of 'old' soil water to the flood hydrograph.
(4) A further unresolved question relates to the linkage between small tributaries and the larger drainage basin, both in terms of flood runoff and the quality of that water. Flood response may well be sensitive to small changes in land use, and the specific location of such changes may also be important. The use of distributed runoff models for studying such linkages is advocated.

REFERENCES

Abdul, A. S., and Gillham, R. W. (1984). Laboratory studies of the effects of the capillary fringe on streamflow generation, *Water Resources Research,* **20**, 691–98.

Acreman, M. C. (1985). Effects of afforestation on the flood hydrology of the upper Ettrick valley, *Scottish Forestry,* **39**, 89–99.

Anderson, M. G., and Burt, T. P. (1978). The role of topography in controlling throughflow generation, *Earth Surface Processes,* **3**, 331–44.

Anderson, M. G., and Burt, T. P. (1982). The contribution of throughflow to storm runoff; an evaluation of a chemical mixing model, *Earth Surface Processes and Landforms,* **7**, 565–74.

Betson, R. P. (1964). What is watershed runoff? *Journal of Geophysical Research,* **69**, 1541–52.

Beven, K. (1982). On subsurface stormflow: an analysis of response times, *Hydrological Sciences Journal,* **27**, 505–21.

Beven, K. (1985). Distributed models, in M. G. Anderson and T. P. Burt, (eds), *Hydrological Forecasting*, Wiley, pp. 405-36.

Beven, K., and Germann, P. F. (1981). Water flow in macropores. II: A combined flow model, *Journal of Soil Science*, **32**, 15-29.

Beven, K. and Germann, P. F. (1982). Macropores and water flow in soils, *Water Resources Research*, **18**, 1311-25.

Boardman, J. (1983). Soil erosion at Aldbourne, West Sussex, England, *Applied Geography*, **3**, 317-29.

Bonnel, M., Hendriks, M. R., Imeson, A. C., and Hazelhoff, L. (1984). The generation of stormflow runoff in a forested clayey drainage basin in Luxembourg. *Journal of Hydrology*, **71**, 53-77.

Bryan, R. B., and Campbell, I. A. (1982). Surface flow and erosion processes in semi-arid meso-scale channels and drainage basins, in D. E. Walling, (ed.), *Recent Developments in the Explanation and Prediction of Erosion and Sediment Yield*, IAHS Publication 137, pp. 123-33.

Bryan, R. B., Yair, A., and Hodges, W. K. (1978). Factors controlling the initiation of runoff and piping in Dinosaur Provincial Park badlands, Alberta, Canada, *Zeitschrift für Gemorphologie*, Supplementband **29**, 151-68.

Burt, T. P. (1979). The relationship between throughflow generation and the solute concentration of soil and streamwater, *Earth Surface Processes*, **4**, 257-66.

Burt, T. P. (1984). Slopes and slope processes. *Progress in Physical Geography*, **8**, 570-82.

Burt, T. P. (1986). Runoff processes and solutional denudation rates on humid temperate hillslopes, in S. T. Trudgill (ed.), *Solute Processes*, Wiley, pp. 193-250.

Burt, T. P. (1987). Slopes and slope processes, *Progress in Physical Geography*, **11**, 598-611.

Burt, T. P., and Arkell, B. P. (1986). Variable source areas of stream discharge and their relationship to point and non-point sources of nitrate pollution, *IAHS Publication 157*, pp. 155-164.

Burt, T. P., and Arkell, B. P. (1987). Temporal and spatial patterns of nitrate losses from an agricultural catchment, *Soil Use and Management*, **3**, 138-42.

Burt, T. P., and Butcher, D. P. (1985a). On the generation of delayed peaks in stream discharge, *Journal of Hydrology*, **78**, 361-78.

Burt, T. P., and Butcher, D. P. (1985b). Topographic controls of soil moisture distribution, *Journal of Soil Science*, **36**, 469-76.

Burt, T. P., and Butcher, D. P. (1986). Stimulation from simulation? A teaching model of hillslope hydrology for use on microcomputers, *J. Geography in Higher Education*, **10**, 23-39.

Burt, T. P., Donohoe, M. A., and Vann, A. R. (1983). The effect of forestry drainage operations on upland sediment yields: the results of a storm-based study, *Earth Surface Processes and Landforms*, **8**, 339-46.

Burt, T. P., and Gardiner, A. T. (1984). Runoff and sediment production in a small peat-covered catchment: some preliminary results, in T. P. Burt and D. E. Walling (eds), *Catchment Experiments in Fluvial Geomorphology*, Geobooks, Norwich, pp. 133-52.

Chorley, R. J. (1978). The hillslope hydrological cycle, in M. J. Kirkby, (ed.), *Hillslope Hydrology*, Wiley, pp. 1-42.

Colborne, G. J. N., and Staines, S. J. (1985). Soil erosion in south Somerset, *Journal of the Agricultural Society*, Cambridge, **104**, 107-12.

Coles, N., and Trudgill, S. T. (1985). The movement of nitrate fertilizer from the soil surface to drainage waters by preferential flow in weakly structured soils, Slapton, south Devon, *Agriculture, Ecosystems and Environment*, **13**, 241-59.

Dietrich, W. E., and Dunne, T. (1978). Sediment budget for a small catchment in mountainous terrain, *Zeitschrift für Geomorphologie*, Supplementband **29**, 191-206.

Dietrich, W. E., Wilson, C. J., and Reneau, S. L. (1986). Hollows, colluvium and landslides in soil-mantled landscapes, in A. D. Abrahams, (ed.), *Hillslope Processes*, Allen and Unwin, pp. 361-88.

De Ploey, J. (1984). Hydraulics of runoff and loess loan deposit, *Earth Surface Processes and Landforms*, **9**, 533-40.

Dunne, T. (1978). Field studies of hillslope flow processes, in M. J. Kirkby, (ed.), *Hillslope Hydrology*, Chichester, Wiley, pp. 227-93.

Dunne, T. (1980). Formation and Controls of Channel Networks, *Progress in physical geography*, **14**, 211-239.

Dunne, T., and Aubry, B. F. (1986). Evaluation of Horton's theory of sheetwash and rill erosion on the basis of field experiments, in A. D. Abrahams, (ed.), *Hillslope Processes*, Allen and Unwin, pp. 31-54.

Dunne, T., and Black, R. D. (1970). An experimental investigation of runoff production in permeable soils, *Water Resources Research*, **6**, 478-90.

Emmett, W. W. (1970). *The hydraulics of overland flow on hillslopes*, U.S. Geological Survey Prof. Paper 662-A, 68 pp.

Freeze, R. A. (1986). Modelling interrelationships between climate, hydrology and hydrogeology and the development of slopes, in M. G. Anderson and K. S. Richards, (eds), *Slope Stability*, Wiley, pp. 381-404.

Fullen, M. A., and Harrison-Reed, A. (1986). Rainfall, runoff and erosion on bare arable soils in East Shropshire, England, *Earth Surface Processes and Landforms*, **11**, 413-25.

Germann, P. F. (1986). Rapid drainage response to precipitation, *Hydrological Processes*, **1**, 3-14.

Gillham, R. W., and Abdul, A. S. (1986). Reply. *Water Resources Research*, **22**, 839.

Gilman, K., and Newson, M. D. (1980). *Soil pipes and pipeflow: a hydrological study in upland Wales*, British Geomorphological Research Group Research Monograph No. 1, Geobooks, Norwich, 110 pp.

Hewlett, J. D. (1961). *Watershed management*, Report for 1961, South Eastern Forest Experimental Station, U.S. Forest Service, ASheville, North Carolina.

Hewlett, J. D. (1982). *Principles of Forest Hydrology*, University of Georgia Press, Athens.

Hewlett, J. D., Cunningham, G. B., and Troendle, C. A. (1977). Predicting stormflow and peakflow from small basins in humid areas by the R-index method, *Water Resources Research*, **13**, 231-53.

Hewlett, J. D., and Helvey, J. D. (1970). Effects of forest-felling on the storm hydrograph. *Water Resources Research*, **6**, 768-82.

Hewlett, J. D., and Hibbert, A. R. (1967). Factors affecting the response of small watersheds to precipitation in humid areas, in W. E. Sopper and H. W. Lull, (eds), *Proceedings of the International Symposium of Forest Hydrology*, Geobooks, Norwich, 275-90.

Horton, R. E. (1933). The role of infiltration in the hydrological cycle, *Transactions of the American Geophysical Union*, **14**, 446-60.

Horton, R. E. (1945). Erosion development of streams and their drainage basins; hydrophysical approach to quantitative morphology, *Bulletin of the Geological Society of America*, **56**, 275-370.

Hursh, C. R., and Brater, E. F. (1941). Separating storm hydrographs from small drainage areas into surface and subsurface flow, *Transactions of the American Geophysical Union*, 863-70.

Imeson, A. C., Vis, M., and Duysings, J. J. H. M. (1984). Surface and subsurface sources of suspended solids in forest drainage basins in the Kemper region of Luxembourg, in

T. P. Burt and D. E. Walling, (eds), *Catchment Experiments In Fluvial Geomorphology*, Geobooks, Norwich, pp. 219–34.

Jones, J. A. A. (1979). Extending the Hewlett model of stream runoff generation. *Area*, **11**, 110–14.

Jones, J. A. A. (1981). *The nature of soil piping: a review of research*. British Geomorphological Research Group Research Monograph No. 3, Geobooks, Norwich.

Jones, J. A. A. (1987). The initiation of natural drainage networks, *Progress in Physical Geography*, **11**, 598–611.

Kennedy, V. C., Kendall, C., Zellweger, G. W., Wyermand, T. A., and Avazino, R. J. (1986). Determination of the components of stormflow using water chemistry and environmental isotopes, Mattole River basin, California. *Journal of Hydrology*, **84**, 107–40.

Kirkby, M. J. (1978). Implications for sediment transport, in M. J. Kirkby, (ed.), *Hillslope Hydrology*, Wiley, pp. 325–63.

Kirkby, M. J., and Chorley, R. J. (1967). Throughflow, overland flow and erosion, *Bulletin of the International Association of Scientific Hydrology*, **12**, 5–21.

Kneale, W. R. (1986). The hydrology of a sloping, structured clay soil at Wytham, near Oxford, England, *Journal of Hydrology*, **85**, 1–14.

Kneale, W. R. and White, R. E. (1984). The movement of water through cores of a dry (cracked) clay-loam grassland topsoil, *Journal of Hydrology*, **64**, 361–5.

Langan, S. J. (1986). *Atmospheric deposition, afforestation and water quality at Loch Dee, S.W. Scotland*, unpublished Ph.D. thesis, University of St Andrews.

Lynch, J. A., Hanna, C. M., and Corbett, E. J. (1986). Predicting pH, alkalinity, and total acidity in stream water during episodic events, *Water Resources Research*, **22**, 905–12.

Meyer, L. D. (1986). Erosion processes and sediment properties for agricultural cropland, in A. D. Abrahams, (ed.), *Hillslope Processes*, Allen and Unwin, pp. 55–76.

Moore, I. D., Burch, G. J., and O'Loughlin, E. M. (1986). Comments on 'Soil erosion class and landscape position,' *Journal of the Soil Science Society of America*, **50**, 1374–5.

Mosley, M. P. (1979). Streamflow generation in a forested watershed, New Zealand, *Water Resources Research*, **15**, 795–806.

National Environment Research Council (1975). *Flood Studies Report*, London (5 volumes).

Newson, M. D. (1980). The geomorphic effectiveness of floods—a contribution stimulated by two recent events in mid-Wales, *Earth Surface Processes*, **5**, 1–16.

O'Loughlin, E. M. (1981). Saturated regions in catchments and their relations to soil and topographic properties, *Journal of Hydrology*, **53**, 229–46.

O'Loughlin, E. M. (1986). Prediction of surface and saturated zones in natural catchments by topographic analysis, *Water Resources Research*, **22**, 794–804.

Pearce, A. J., Stewart, M. K., and Sklash, M. G. (1986). Storm runoff generation in humid headwater catchments, 1. Where does the water come from? *Water Resources Research*, **22**, 1263–272.

Philip, J. R. (1957). The theory of infiltration: 4. Sorptivity and algebraic infiltration equations, *Soil Science*, **84**, 257–64.

Pilgrim, D. H., Huff, D. D., and Steele, T. D. (1979). Use of specific conductance and contact time relations for separating storm flow components in storm runoff, *Water Resources Research*, **15**, 329–39.

Reid, I., and Parkinson, R. J. (1984a). The wetting and drying of a grazed and ungrazed clay soil, *Journal of Soil Science*, **35**, 607–14.

Reid, I., and Parkinson, R. J. (1984b). The nature of the tile-drain outfall hydrograph in heavy clay soils, *Journal of Hydrology*, **72**, 289–305.

Reynolds, B., Neal, C., Hornung, M., and Stevens, P. A. (1986). Baseflow buffering of streamflow acidity in five mid-Wales catchments, *Journal of Hydrology*, **87**, 167–85.
Robinson, M. (1986). Changes in catchment runoff following drainage and afforestation, *Journal of Hydrology*, **86**, 71–84.
Robinson, M. and Beven, K. J. (1983). The effect of mole drainage on the hydrological response of a swelling clay soil, *Journal of Hydrology*, **64**, 205–23.
Sherman, L. K. (1932). Streamflow from rainfall by the unit hydrograph method, *Engineering News Record*, **108**, 501–5.
Sklash, M. G., and Farvolden, R. N. (1979). The role of groundwater in storm runoff, *Journal of Hydrology*, **43**, 45–65.
Thorne, C. R., Zevenbergen, L. W., Burt, T. P. and Butcher, D. P. (1987). Terrain analysis for quantitative description of zero order basins, in R. L. Beschta, T. Blinn, G. E. Grant, G. G. Ice, and F. J. Swanson, (eds), *Erosion and Sedimentation in the Pacific Rim*, IAHS publication No. 165, pp. 121–30.
Trudgill, S. T. (1986). Soil water dye tracing, with special reference to the use of Rhodamine WT, Lissamine ff and Amino G Acid, *Hydrological Processes*, **1**, 149–70.
Tsukamoto, Y., Ohta, T., and Noguchi, H. (1982). Hydrological and geomorphological studies of debris slides of forested hillslopes in Japan, in D. E. Walling, (ed.), *Recent Developments in the Explanation and Prediction of Erosion and Sediment Yield*, IAHS Publication 137, pp. 89–98.
Walling, D. E. (1983). The sediment delivery problem, *Journal of Hydrology*, **65**, 209–37.
Walling, D. E., and Webb, B. W. (1975). Spatial variation of river water quality; a survey of the River Exe. *Transactions of the Institute of British Geographers*, **65**, 155–71.
Walling, D. E. and Webb, B. W. (1980). The spatial dimension in the interpretation of stream and solute behaviour, *J. Hydrology*, **47**, 129–49.
Warren, S. C. (1986). *Acidity in United Kingdom Fresh Waters*. UK Acid Waters Review Group, Department of the Environment, Crown Copyright.
Welsh, W. T., Tervet, D. J., and Hutchinson, P. (1986). *Acidification Investigations in South-west Scotland*, unpublished report to Solway River Purification Board, Scotland.
Whipkey, R. Z. (1965). Subsurface stormflow from forested watersheds, *Bulletin of the International Association of Scottish Hydrology*, **10**, 74–85.
Willis, G. K., and McDowell, L. L. (1982). Pesticides in agricultural runoff and their effects on downstream water quality, *Environmental Toxicology and Chemistry*, **1**, 267–9.
Wischmeier, W. H., and Smith, D. D. (1965). Predicting rainfall-erosion losses from cropland east of the Rocky Mountains, *USDA Agricultural Handbook*, 282 pp.
Wischmeier, W. H., and Smith, D. D. (1978). *Predicting Rainfall Erosion Losses–A Guide to Conservation Planning*, USDA—SEA Agriculture Handbook, 537 pp.
Yair, A., and Lavee, H. (1985). Runoff generation in arid and semi-arid zones, in M. G. Anderson and T. P. Burt, (eds), *Hydrological Forecasting*, Wiley, pp. 183–220.
Yair, A., Lavee, H., Bryan, R. B., and Adar, E. (1980). Runoff and erosion processes and rates in the Zin valley badlands, northern Negev, Israel, *Earth Surface Processes*, **5**, 205–23.
Zaslavsky, D. (1970). *Some Aspects of Watershed Hydrology*, ASDA Research Paper, ARS 41-157, 96 pp.
Zaslavsky, D. and Sinai, G. (1981). Surface Hydrology: 5 parts, *J. Hydraul. Div. Proc. Am. Soc. Civ. Eng.*, **107**, 1–93.

3 Flood Wave Attenuation due to Channel and Floodplain Storage and Effects on Flood Frequency

DAVID R. ARCHER
Northumbrian Water

INTRODUCTION

Most engineering hydrologists have adopted methods described in the *Flood Studies Report* (National Environment Research Council, 1975) as the basis for estimation of design flood discharges at ungauged sites in Britain. Two alternative methods were described, a statistical method and a rainfall–runoff method.

Estimation by the statistical method is carried out in two stages. First the mean annual flood ($\bar{Q}$) is estimated by means of a multiple linear regression of $\bar{Q}$ on catchment characteristics. Secondly, a regional growth curve is used as a multiplier of $\bar{Q}$ to estimate floods of required return period.

The adopted catchment characteristics for the estimation of $\bar{Q}$ were area, channel slope, stream frequency, soil class, short duration net rainfall and percentage of the area urbanised or passing through lakes. These characteristics were chosen for predictive power and independence of each other (as far as possible). Ease of measurement was also considered important and all the characteristics are measured from maps. This is not to suggest that other factors requiring field work to determine are not important. It is the intention of this paper to draw attention to two such related characteristics, channel and floodplain storage. In discussion during the first *Flood Studies Conference* (Institution of Civil Engineers, 1975) Prus-Chacinski drew attention to their omission.

> I am surprised at the omission of another important factor, the width of the flood valley, which affects valley storage. ... The author's method is only loosely connected with geomorphology. The history of the river is written in the river's size and other characteristics of the river's regime ... The size of a river and its valley is the integral of all climatic and geophysical factors.

The potential importance of valley storage was acknowledged at that point and also by Beran (1981a), who suggested that in spite of its omission, the equation for $\bar{Q}$ yields a very realistic picture of the progress of a flood wave because of the compensating interaction of catchment area with stream frequency and channel slope. Nevertheless an explicit floodplain index was thought preferable as this would allow the exponents of other characteristics to adopt more natural values.

A more telling demonstration of the influence of channel and floodplain storage is given by Bailey and Bree (1981), who showed that arterial channel improvements in Ireland resulted in an increase in the three-year return period flood of about 60% without significant change in the slope of the flood frequency graph. It is possible that such major differences occur between natural catchments which are not fully compensated by other variables.

In this study, flood attenuation and the transformation of flood frequency over a common period were investigated over a reach of the lower River Tees in north-east England.

THE RIVER TEES AND THE STUDY REACH

The investigated reach of the River Tees lies between the gauging stations at Broken Scar near Darlington and that at Low Moor (Figure 3.1), 3.4 km upstream from the tidal limit. They have a common gauged record from 1969 to 1986, although the Broken Scar record extends back to 1956.

Both gauging stations have well-defined flood ratings. Broken Scar has a compound Crump weir, modular to near the highest observed flow and confirmed by current meter gauging up to the sixth highest observed level. At Low Moor there is a natural channel control at high flows defined by current meter over 92% of the observed range of level.

The catchment area to Broken Scar is 818 km^2 and that to Low Moor is 1264 km^2. There are two principal tributary catchments to the reach, the Skerne presently gauged at South Park (250 km^2) and the Clow Beck, gauged at Croft until 1980 (78 km^2). Both these tributaries join the Tees near the upstream end of the reach.

The Tees above Broken Scar is an upland river rising on the Pennine plateau with significant areas over 500 m OD and with a catchment average rainfall of 1207 mm. The predominant peat cover has a high degree of impermeability and percentage runoff in flood events commonly exceeds 60%. The main channel in this upper and middle reach is steeply sloping with a number of rapids and falls, and runoff response is rapid. There is little overbank storage upstream from Broken Scar even in extreme flows.

Below Broken Scar there is a marked change both in the channel and in the characteristics of the contributing catchments. The channel length from Broken Scar to Low Moor is 34.6 km and the average slope is 0.92 m/km. Bankfull

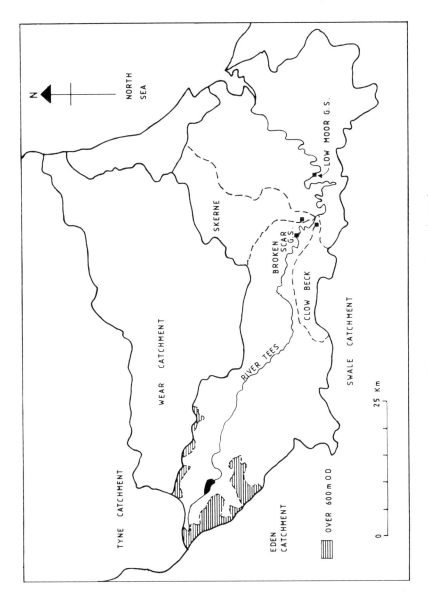

Figure 3.1 The River Tees and the study reach.

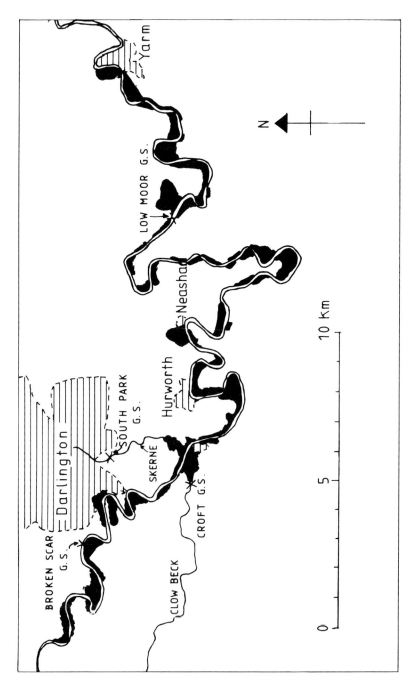

Figure 3.2 River Tees between Broken Scar and Low Moor gauging stations showing extent of overbank flooding in March 1968.

discharge is approximately 350 m³/s and the channel capacity to bankfull through the reach is about 7 Mm³. The alluvial channel meanders over a wide flood plain. The extent of overbank flooding in the largest observed event in March 1968 is shown in Figure 3.2. The catchment contributing to this reach generally lies below 100 m OD and has an average annual rainfall of 665 mm. With more permeable agricultural soils, losses are much higher. Consequently the contribution to flood flows is small in comparison to the flow originating in the Tees above Broken Scar—generally less than 10%.

The combination of a long reach and little intervening inflow provides an ideal opportunity for the study of flood wave movement and attenuation and the transformation of flood frequency.

FLOOD WAVE TRAVEL TIME

The time of travel of the peak discharge from Broken Scar to Low Moor was determined for 64 events at discharges ranging from 20 to 700 m³/s. The relationship of travel time and wave speed to discharge is shown in Figure 3.3. Discharge at Broken Scar was used, since for practical forecasting it is the upstream discharge only which is available. The plotted lines represent smoothed 10, 50, and 90 percentiles of grouped data. Although there is a scatter of points, a clear relationship is observed. Wave speed increases from about

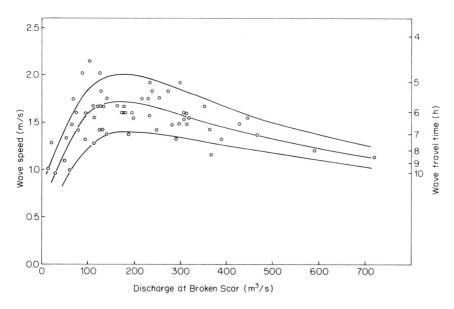

Figure 3.3 Wave speed and travel time through the reach at flood discharges.

1 m/s at 40 m³/s to 1.7 m/s at 120 m³/s. It declines very slightly to about 300 m³/s. Thereafter there is a more distinct reduction in speed through the observed range of discharge. The equivalent minimum wave travel time is approximately 5.6 hours but there is an increase to 8.7 hours at the highest observed flow.

In this analysis the tributary inflow from the River Skerne and other sources has not been considered. The small volume from these is unlikely to affect the estimated travel time seriously, though it may account for some of the spread in the relationship.

FLOOD PEAK ATTENUATION

The relationship between peak inflow to the reach and peak outflow is shown in Figure 3.4. The total peak inflow to the reach was calculated as the sum of peak flow at Broken Scar and the concurrent flow at South Park on the Skerne grossed up for the total tributary catchment.

The graph shows that the degree of attenuation is highly variable over the range of inflow discharges and also at given inflow discharge. Broadly attenuation is at a minimum and nearly zero at about 375^3 m/s (just above bankfull discharge). It increases both for lower discharges contained within the

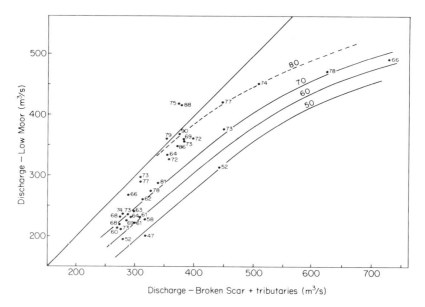

Figure 3.4 Relationship between upstream and downstream discharges with the ratio of 12-hour mean flow to the peak flow as a parameter.

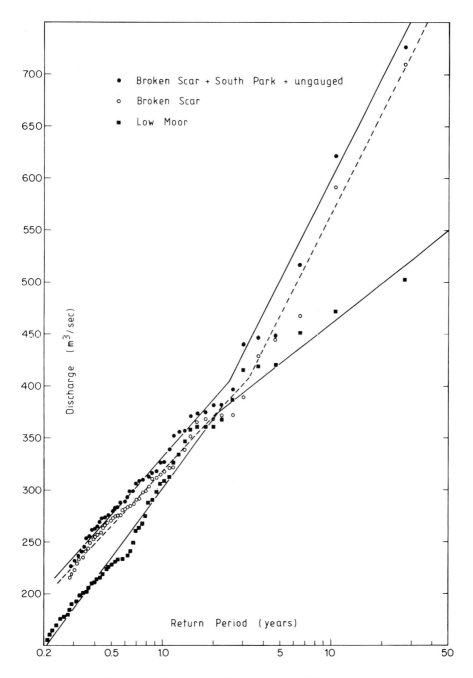

Figure 3.5 Upstream and downstream flood frequency.

channel and also very considerably for inflow discharges above bankfull with increasing incursion on the floodplain.

The effect of flood volume (or peakedness) on attenuation was investigated. Peakedness of the flood inflow to the reach may be represented by the ratio of the 12-hour mean flow to the peak flow at Broken Scar. This parameter is shown for individual events in Figure 3.4 and best fit lines of equal ratio have been drawn by eye. As might be expected the degree of attenuation at a given peak inflow appears to be far greater for sharply peaked low volume floods.

TRANSFORMATION OF FLOOD FREQUENCY

The variable attenuation of flood peaks through the reach is reflected in the comparative upstream and downstream flood frequencies (Figure 3.5). Peaks over a threshold series (POT) were used in preference to annual maxima to permit comparison of more frequent inbank as well as overbank floods. Frequency plots are shown in Figure 3.5 for:

(1) Broken Scar
(2) Broken Scar plus tributary inflow
(3) Low Moor.

The Blom plotting position was used and the lines are eye-guided best-fit to the exponential distribution. However, the conclusions drawn are independent of the choice of distribution or plotting position.

The plot shows convergence of upstream and downstream frequencies up to a return period of 2.2 years. At this return period, discharges at Broken Scar and Low Moor are equivalent at 375 m^3/s whilst the total inflow is about 20 m^3/s higher. At a 0.33-year return period (3 per year frequency) peak inflow exceeds outflow by 55 m^3/s. The plots diverge again at higher return periods and at a 20-year return period the difference is nearly 200 m^3/s.

Flood peaks do not always have the same rank at upstream and downstream stations and therefore points plotted at the same return period may represent different events. An extreme example is the sharply peaked event of 2 January 1976 with a discharge of 429 m^3/s at Broken Scar (Rank 5) which reduced to 313 m^3/s at Low Moor (Rank 15).

DISCUSSION

Archer (1981) suggested that variations in shape and slope of flood frequency distributions in north-east England could be accounted for amongst other things by the characteristics of channel and flood plain storage. This study provides evidence of the nature of the control.

Above bankfull, increasing storage on the floodplain and associated delays due to frictional resistance suppress flood growth at the downstream station. However, mean annual flood is virtually unaffected by overbank flow which does not commence until flow exceeds 350 m^3/s (*circa* 1.7-year return period). The effects of channel storage are seen below bankfull. As flood peak and volume increase, attenuation decreases and downstream flood growth becomes steeper. The resulting flood frequency has a break of slope at bankfull discharge.

Beran (1981b) indicated that observed variations in slope and shape could be accounted for by random sampling from an exponential distribution of POT values. However, in this instance the use of dependent data sets enables the effect of catchment and channel to be distinguished from the climatic input. It seems likely that transformation in the reach would give the observed form of downstream distribution for a range of upstream distributions.

This transformation has important implications for flood estimation at ungauged sites. Floodplain storage and attenuation is a common phenomenon and it is suggested that suppression of flood growth may be similarly widespread. Indeed several other catchments in north-east England with significant floodplain storage have supressed flood growth and a break of slope around bankfull stage. However, the effect may be obscured in less ideal reaches where there is a larger lateral inflow contribution, and this could cause difficulties in extrapolation to other sites.

Where flood growth curves are based on a defined geographical region it is inconvenient to separate out those with significant floodplain storage. Alternative methods of classifying basins into homogeneous groups using cluster analysis (Acreman and Sinclair, 1986) or interactive search techniques to optimise the grouping efficiency (Wiltshire, 1986) offer more opportunity for the inclusion of floodplain and channel effects, but this has not yet been done.

There are also practical consequences of confinement of flood flows within flood banks through the reach. The effects of reduced attenuation and shorter flood travel time on property at risk downstream of the protected reach need evaluation.

ACKNOWLEDGEMENTS

This paper is published with the permission of Mr N. Ruffle, Director of Operations at Northumbrian Water.

REFERENCES

Acreman, M. C., and Sinclair, C. D. (1986). Classification of drainage basins according to their physical characteristics; an application for flood frequency analysis in Scotland, *J. Hydrol.*, **84**, 365–80.

Archer, D. R. (1981). A catchment approach to flood estimation, *J. Instn. Wat. Engrs. Sci.*, **35**(3), 275–89.

Bailey, M. A., and Bree, T. (1981). Effect of improved land drainage on river flood flows, in *Flood Studies Report–Five Years On*, Thomas Telford, London, pp. 131–42.

Beran, M. A. (1981a). Recent advances in statistical flood estimation techniques, in *Flood Studies Report–Five Years On*, Thomas Telford, London, pp. 25–32.

Beran, M. A. (1981b). Communication on A catchment approach to flood estimation, *J. Instn. Wat. Engrs. Sci.*, **35**(3), 528–35.

Institution of Civil Engineers (1975). *Flood Studies Conference*. Discussion on Papers 7 and 8, ICE, London, p. 61.

Natural Environment Research Council (1975). *Flood Studies Report*, London (5 volumes).

Wiltshire, S. E. (1986). Identification of homogeneous regions for flood frequency analysis, *J. Hydrol.*, **84**, 287–302.

Floods: Hydrological, Sedimentological and Geomorphological Implications
Edited by K. Beven and P. Carling
© 1989 John Wiley & Sons Ltd

4 Physically Based Hydrological Models for Flood Computations

SÁNDOR AMBRUS
Water Resources Research Centre, P.O.B. 27 H-1453 Budapest

LÁSZLÓ IRITZ
Uppsala University, Division of Hydrology, V. Ågatan 24 S-752 20 Uppsala

ANDRÁS SZÖLLÖSI-NAGY
Water Resources Research Centre, P.O.B. 27 H-1453 Budapest

FLOOD PROTECTION AND CONTROL

The decision-making procedure during floods on big rivers is primarily based on hydrological computations. The proposed flow-chart of the decision-making organization can be seen in Figure 4.1.

Information on the state of the watershed is continuously forwarded to the centre. Data processing, checking and completion are done and forecasts are made from the latest data set. On the basis of this information, the actual state of the watershed can be analysed and, if it is needed, an alarm signal for flood alert is given. On the basis of the predictions made from one cross-section to another, the surface profile for the longitudinal section of the whole reach has to be predicted, too. From the predicted longitudinal surface profile, the expected load on the levee system can be estimated. The predicted peak levels increased by a safety margin can then be compared with the multiannual maximum and the design peak level. As a next step—in the case of an emergency—recommendations are made for the decisions about the means of flood control to be executed. Then follows a preliminary technical analysis and a preliminary cost–benefit analysis of the possible options as well as an estimation of their outcomes. The effects of the technical analysis are continuously observed and, as much as possible, subject to control (as e.g. the opening of levees and flood release basins).

Operational practice needs methods with limited data requirements and fast adaptation possibilities. The simplified hydraulic and hydrological models are designed to respond to these demands. In the following, recently developed flood control models will be presented.

48 Floods

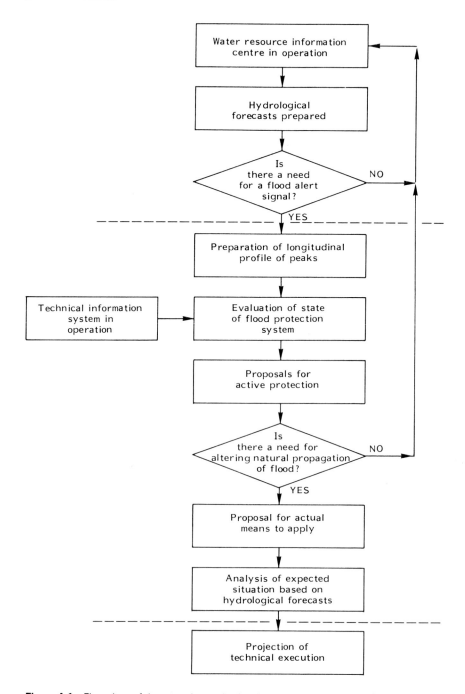

Figure 4.1 Flow chart of the central organization that supports operational flood protection.

DISCRETE CASCADE MODEL FOR THE APPROXIMATE UNSTEADY FLOW EQUATIONS

A first-order approximation of the Saint-Venant equations, describing the gradually varying unsteady flow in an open channel, was given by Kalinin and Milyukov (1958) who introduced the notion of *characteristic reach*. The characteristic reach technique replaces the equation of moments by a linear relationship between channel storage and outflow from the characteristic reach. The solution is formally identical with that of the Nash cascade (1957) used for rainfall/runoff modelling. Both continuous models are linear lumped time-invariant representations.

The continuous model cannot be directly applied in practice since

(1) measurement data are obtained at discrete time epochs and
(2) data processing is done on digital computers and therefore a discrete model is required.

Szöllösi-Nagy (1982) has shown that if the continuous input/output functions are sampled at time intervals Δt apart, then the discrete state-space model

$$\mathbf{x}_{t+\Delta t} = \varphi(\Delta t)\mathbf{x}_t + \Gamma(\Delta t)u_t \tag{4.1}$$

$$y_t = \mathbf{H}\mathbf{x}_t, \tag{4.2}$$

where $u(.)$ is the inflow (input) to the cascade containing n elemental linear reservoirs/characteristic reaches; $x(.)$ is the state vector consisting of the storage of the linear elements; and $y(.)$ is the outflow from (output of) the linear cascade. The dynamic system of the linear cascade is characterized by the matrix triplet $\Sigma_D = (\varphi, \Gamma, \mathbf{H})$:

$$[\varphi(\Delta t)]_{i,j} = \begin{cases} \dfrac{(k\Delta t)^{i-j}}{(i-j)!} e^{-k\Delta t}, & i \geq j \\ 0, & i \leq j \end{cases} \tag{4.3}$$

is the state-transition matrix and

$$[\Gamma(t)]_i = (1-e^{-k\Delta t} \sum_{j=0}^{i-1} \frac{(k\Delta t)^j}{j!})/k \tag{4.4}$$

$$\mathbf{H} = [0, 0, \ldots, k], \tag{4.5}$$

where $k = 1/k$ is the travel time/storage coefficient.

The discrete model defined by equations (4.1) and (4.2) is operationally used for real-time forecasting of river flow (Szöllösi-Nagy 1986).

The exact timing of opening the emergency reservoir will be based on a recently developed algorithm for streamflow input detection (Szöllösi-Nagy, 1986). Figure 4.2 shows the sketch of an emergency reservoir and its environment and also the problem to be solved. The summed-up difference between the two (predicted and allowed) flood waves has to be stored in the emergency reservoir. The input detection is practically the inverse problem of downstream forecasting and can also be solved by the discrete cascade model, (Szöllösi-Nagy and Iritz, 1985).

The exact solution is based on the assumption that the model parameters k and n are known as well as the initial state vector x_0 and the output series, y_τ for $\tau = 1, 2 \ldots, t + 1 > n$. From equations (4.1) and (4.2) it follows that

$$y_{t+1} = \mathbf{H}\,\mathbf{x}_{t+1} \tag{4.6}$$

$$y_{t+1} = \mathbf{H}\varphi\mathbf{x}_t + h_1 u_t,$$

where

$$h_1 = \mathbf{H}\,\Gamma = 1 - e^{-k} \sum_{i=0}^{n=1} \frac{k^i}{i!} \tag{4.7}$$

is the first value of the impulse response of the discrete system. The value of h_1 has to be positive. The input at time t can be expressed from equation (6):

$$u(t) = \frac{1}{h_1}(y_{t+1} - \mathbf{H}\varphi\mathbf{x}_t) \tag{4.8}$$

This is a recursive formula for the computation of the input series at time $t = 0, 1, 2, \ldots, n$. The recursion starts at time $t = 0$ using the previously (off-line) determined initial state vector. Only the first n input–output data pairs are needed. Input detection begins at time n, and, since x_0 and the series $u_0, u_1, \ldots, u_{n-1}$ are known, the states of the cascades can be computed from equation (4.6) as:

$$\mathbf{x}_n = \varphi\mathbf{x}_{n-1} + \Gamma\,\mathbf{u}_{n-1}. \tag{4.9}$$

From the known output y_{n+1} the input u_n can be detected from equation (4.8). In the next step, $\mathbf{x}_{n+1}$ is generated, and this uniquely determined (together with Y_{n+2}) the input u_{n+1}, and so forth. The observability of the discrete cascade is a necessary and sufficient condition of detectability.

Physically Based Hydrological Models for Flood Computations

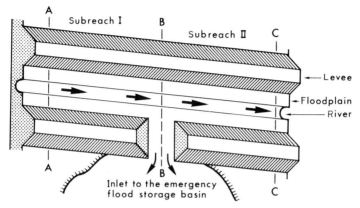

The inlet of the emergency flood release basin

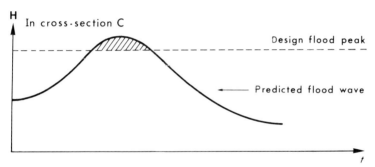

A flood wave that exceeds design peak level in cross-section C

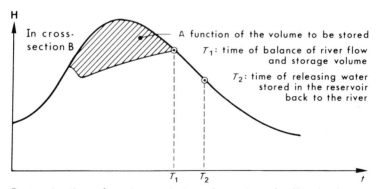

Determination of optimum timing of opening the flood release reservoir

Figure 4.2 Sketch of an emergency flood storage basin and the problem to be solved in connection with flood regulation.

A JOINT APPLICATION OF THE EQUATIONS OF THE KINEMATIC AND DIFFUSION WAVES

In the above models a simple state-space description of the kinematic wave has been utilized. Another approach to the phenomenon of flood modelling is to use a higher-order approximation of the Saint-Venant equations containing the diffusion term. The equations can be brought to the form of a single differential equation of second order, after Dooge (1973):

$$\frac{\partial Q}{\partial t} = D(Q)\frac{\partial^2 Q}{\partial x^2} - C(Q)\frac{\partial Q}{\partial x} \qquad (4.10)$$

where $C(Q)$ and $D(Q)$ are nonlinear discharge-dependent coefficients that make no explicit solution to the equation possible.

The first term on the right-hand side of equation (4.10) is the diffusion term, which is actually a quantity one magnitude smaller than the runoff velocity and affects the attenuation of the flood wave. It attenuates the flood wave in both directions about the peak, as is superimposing little waves or impulses, travelling in both directions, upstream and downstream, away from the points of observation.

Actually, a break or sudden opening of a gate on the levee system can induce a similar type of phenomenon: a 'negative' impulse to the water motion at the breaking point of the reach considered, acting again in both directions, upstream and downstream. The equation of the diffusion wave is therefore very well suited for the description of the abrupt change and its further effects in the water regime after the opening of a gate to a flood release basin.

However, there is a difference in the magnitude of a diffusion and the dambreak related to the discharge of a river: the latter one has a much more dominant impact on the water regime: it has to be at least the same order of magnitude as the runoff discharge itself. Thus, the second-order term in equation (4.10) has to be replaced by one of order 1.

The system description of the model and even the system matrix remain thus formally the same, only its terms in equation (4.1) will be somewhat different (Ambrus and Iritz, 1985):

$$\frac{d}{dt}x(t) = \begin{bmatrix} -\varphi_0 & \varphi_2 & & & \\ \varphi_1 & \varphi_0 & & & \\ & \cdot & \cdot & \cdot & \\ & & \cdot & \cdot & \varphi_2 \\ & & & \varphi_1 & \varphi_0 \end{bmatrix} \begin{bmatrix} y_{1t} \\ y_{2t} \\ \cdot \\ \cdot \\ y_{nt} \end{bmatrix} + \begin{bmatrix} \varphi_1 & 0 \\ 0 & 0 \\ \cdot & \cdot \\ \cdot & \cdot \\ 0 & \varphi_2 \end{bmatrix} \begin{bmatrix} y_{0t} \\ y_{n+1,t} \end{bmatrix} \qquad (4.11)$$

where

$$\varphi_0 = \frac{C_1}{\Delta x} + 2\frac{C_2}{\Delta x}; \quad \varphi_1 = \frac{C_1}{\Delta x} + \frac{C_2}{\Delta x}; \quad \varphi_2 = \frac{C_2}{\Delta x}$$

where C_1 is discharge velocity and C_2 the motion velocity of the impulse, which in our case (since it is an intake) has negative sign related to the streamflow. In this kind of description, the impulse moves along the whole reach and is proportional to the water quantity of the cross-sections.

The diffusion wave model will be applied for the interpolation of the longitudinal surface between the three reference cross-sections of the reach.

CONCLUSIONS

Numerical tests have been made for the Körös River basin, which is of great importance for flood control protection in Hungary. Figures 4.3–4.5 present the results of the computations for the high flood in 1980, when the emergency storage basin was in use (Harkányi and Szöllösi-Nagy, 1983).

Figure 4.3 shows the forecasted and observed discharges for the whole year.

Figure 4.4 presents two hydrographs together. The continuous line was the hydrograph in cross-section A (see Figure 4.2) of the upstream station (Malomfok) and the dotted line is the computed hydrograph in the river cross-section B at the outlet into the emergency flood storage basin (called Mályvád).

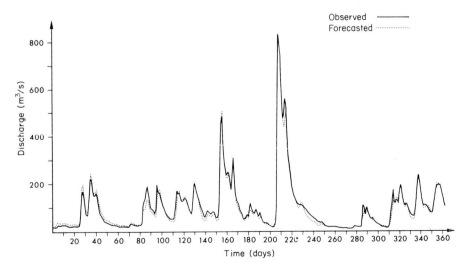

Figure 4.3 Forecasted and observed discharges in 1980 on the River Körös at the station Bekes.

54   Floods

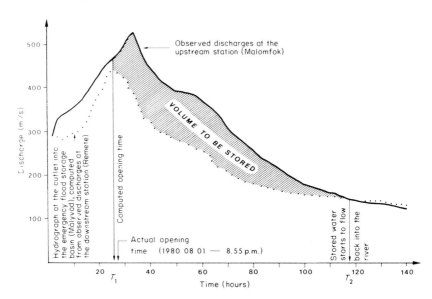

Figure 4.4 Flood storage simulation for River Körös by input detection.

The computation was performed on time series in the cross-section C of the downstream station (Remete) for the period during which the emergency storage basin was open. The peak of the computed hydrograph indicates the opening time of the storage basin (T_1). We can also see the moment when the water level in cross-section B and in the storage basin was the same (T_2). When the water

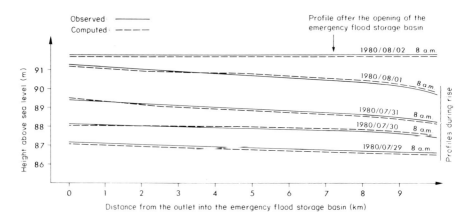

Figure 4.5 Computed and observed longitudinal water profiles on the River Körös, downstream from the opened emergency flood storage basin (Malyvad).

flows back from the storage basin, the discharge in cross-section B became higher than at the upstream station C. The difference between two hydrographs gives the stored water volume.

Figure 4.5 shows the longitudinal water profile computed by the diffusion wave model and the observed level for the same flood regulation event.

Nowadays all three methods operate in the hydrological support system of flood control in Hungary.

REFERENCES

Ambrus, S., and Iritz, L. (1985). *Tasks of Hydrological Forecasting and Hydraulic Computation in Flood Control Operation*, Research Report (T. sz. 7611-3/1/11), VITUKI, Budapest, Baja.

Dooge, J. C. I. (1973). *Linear Theory of Hydrologic Systems*, USDA Techn. Bull. No. 1468. Washington, D.C.

Harkányi, K., and Söllösi-Nagy, A. (1983). Deriving operation rules for emergency flood control reservoirs by a discrete dynamic linear cascade model, *Proc. XX IAHS Congress*, Moscow.

Kalinin, G. P., and Milyukov, P. I. (1958). Pribliezhenyi raschet neustanovivshegosya dizheniya vodnykh mass (Approximate calculation of unsteady flow of water masses), in Russian, *Trudy TsIP*, No. 66, Leningrad.

Nash, I. E. (1957). The form of the instantaneous unit hydrograph, *Proc. IAHS Assembly-Gen.*, 3, 114–31, Toronto, Ontario.

Szöllösi-Nagy, A. (1982). The discretization of the continuous cascade by means of state space analysis, *Journal of Hydrology*, **58**, (3–4).

Szöllösi-Nagy, A. (1986). Hydrological forecasts and warnings, in Ö. Starosolszky (ed.), *Applied Surface Hydrology*, Water Resources Publications, Littleton, Colorado, Part III.

Szöllösi-Nagy, A., and Iritz, L. (1985). *Methods Developed for Flood Forecasting and Control in Complex Flood Protection Systems. Overview for Training Courses of Decision-makers in Flood Protection Control*, Research Report (T.sz. 7611/1/41), VITUKI, Budapest-Baja.

5 Flood Frequency and Urban-Induced Channel Change: Some British Examples

CAROLYN R. ROBERTS
School of Geography and Geology, College of St Paul and St Mary, Cheltenham

INTRODUCTION

Urbanisation is often cited as one of the major human modifications to catchment hydrology in developed areas (UNESCO, 1979), and a series of case studies in the last two decades have attempted to establish how the hydrological change has affected channel morphology in downstream reaches (Wolman, 1967; Hammer, 1972; Leopold, 1973; Graf, 1975; Mosley, 1975; Fox, 1976; Gregory, 1976, 1977; Gregory and Park, 1976; Hollis and Luckett, 1976; Richards and Wood, 1977; Robinson, 1976; Park, 1977; Knight, 1979; Morisawa and LaFlure, 1979; Nanson and Young, 1981; Whipple and Dilouie, 1981; Neller, 1988). Bed and bank erosion, loss of riparian trees and damage to riverbank structures are the most frequently cited changes, with phases of sedimentation in the early stages of development later being followed by flushing out of construction debris and topsoil released from cleared areas. Perhaps partly because of the cautionary notes concerning the methodology used to identify channel change sounded by Thornes (1977) and Richards and Greenhalgh (1984), there has been little systematic attempt to link the magnitude of channel changes to the nature of the urban area, or to establish the role played by floods of different recurrence interval in the adjustment process.

Four further reasons for these omissions are probable. Firstly, few gauging stations have long, accurate records of flow conditions before, during and after substantial urban development (exceptions being development in some of Britain's new towns (Hollis, 1974; Knight, 1979)). Secondly, urban areas are complex, containing a matrix of different land uses, development densities and ages of development, spread across variable proportions of catchments; this makes indirect establishment of the probable magnitude of hydrological change in ungauged catchments problematic. Thirdly, many channels in catchments experiencing substantial development will be culverted, channelised or artificially excavated to increase their floodwater carrying capacity (Brookes,

Table 5.1 Physical parameters of the study areas.

Study area	Total no. sites	No. rural sites	Mean annual rainfall (mm)[1]	Two-day M5 rainfall (mm)[1]	Winter rain acceptance potential[2]	Mean soil moisture deficit (mm)[3]	Catchment geology index[4]
R. Okement, Okehampton	37	23	1600	100	4	4	3
Hawkcombe Stream, Porlock	38	20	1400	75	3	7	2
Wray Brook, Moretonhampstead	40	30	1200	75	2	6	2
Red River, Camborne	64	19	1150	65	2	5	3
Luggie Water and Red Burn, Cumbernauld	85	35	1100	60	4	3	3
River Gissage, Honiton	30	18	1000	70	2	8	3
Savick Brook, Fulwood	45	23	1000	55	4	6	4
River Tawd, Skelmersdale	97	47	950	55	3	6	3
River Bourre, Tidworth	59	23	850	50	1	9	1
Fenay Beck, Huddersfield	63	13	800	60	3	4	3
Avon tributaries, Chippenham	48	25	800	55	3	8	2
Pin Brook, Pinhoe, Exeter	38	18	800	55	2	8	2
River Soar tributaries, Leicester	40	6	650	45	4	11	4
Sow Brook, Rugby	21	5	650	45	1	11	2
Stevenage Brook, Stevenage	124	59	650	45	2	15	2
River Ingrebourne, Harold Hill	42	18	600	45	4	15	4

Note: Climatic and soil information based on *Flood Studies Report* (NERC, 1975).

[1] From observations
[2] 1 = very high, 5 = very low
[3] By calcuation
[4] 1 = very permeable, 4 = very impermeable

1985). Consequently any gradual flow-induced changes are masked. Finally, the natural variability of channel morphology both spatially and temporally poses problems for the would-be identifier of 'significant' channel change. As will be demonstrated, the effects of urbanisation on channel morphology are usually limited in comparison to those of long return period flood events, but the flood frequency characteristics of catchments do appear to play an important part in governing the overall extent of urban-induced change.

Whilst adaptations are possible in a range of channel parameters, this study considers specifically the bankfull cross-sectional area ('capacity') of the channel. Within the studied reaches, adaptations to planform identified from historical data sources were minor, and usually the result of deliberate intervention by drainage or civic authorities. This lack of planform response to urbanisation may be the result of bias in sampling rivers for investigation as most British towns lie in lowland locations, away from the very powerful, mobile, high-gradient, gravel-bed streams of the uplands. The 'feebler' streams of southern and midland Britain (Ferguson, 1981) are thus prominent in the analysis, which centres on sixteen urban areas (Figure 5.1). Alternatively, planform changes may only manifest themselves over longer periods of time than those available following the development of urban areas in Britain. Some characteristics of the study catchments are shown in Table 5.1.

Urban drainage systems and the implications for natural channels

The implications of urban stormwater drainage systems for catchment hydrology have been studied at a variety of scales from a single small paved area, to the level of whole suburbs or towns, although the emphasis has usually reflected engineers' interest in the sizing of pipework (see for example, Helliwell and Kidd, 1979). Commonly, a short length of suburban road with associated buildings has formed the focus of study and the results are of limited value in understanding the hydrological response of urban areas in excess of a few hectares, or in establishing the precise implications of the urbanisation process for any one catchment. Furthermore, the development of urban areas in Britain encapsulates decades when drainage designers favoured combined stormwater and foulwater systems, rather than the separate systems usual today.

The variability of existing British urban drainage systems is summarised in Figure 5.2, which suggests that the extent of hydrological change will be governed by the nature of the connections between river and impermeable or drained areas. Whilst modern suburbs are usually well served by large diameter drains, allowing rapid runoff of rainfall direct to the nearest watercourse, older development often has a combined system in which the majority of both surface and foul water enters the river only after passing through a sewage treatment plant. Hydrological change is hence minimised and it would be unreasonable to expect major channel adjustments to follow. Assuming that suitable connections

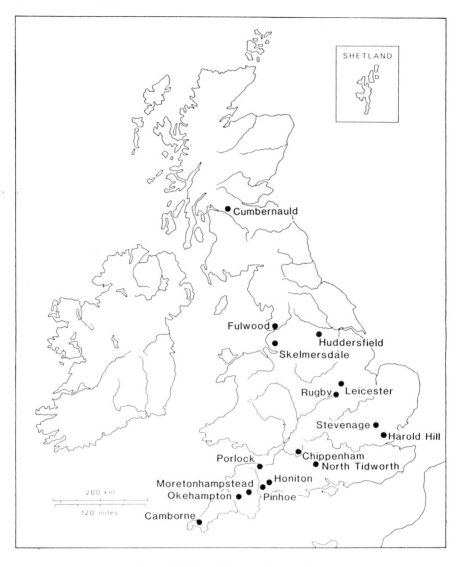

Figure 5.1 Map of selected study sites.

do exist for flood runoff, then urban areas have been shown to increase flood peaks in the channel and to decrease lag times, as well as inducing a variety of other changes in the shape of the hydrograph. It is the magnitude of flood peaks which is of greater significance for controlling channel dimensions. Leopold (1968), Crippen and Waananen (1969) and Hollis (1975) have provided useful summaries of research in this area, showing that the order of magnitude of

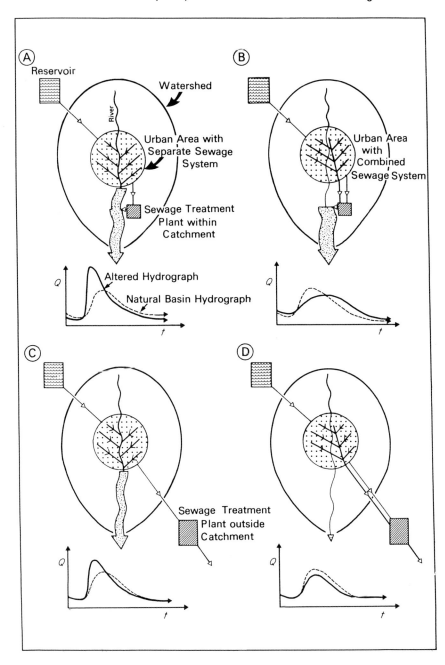

Figure 5.2 Variability of urban hydrological systems in Britain, and the implications for changes in flood hydrographs.

changes in flood peaks is related to the extent of urbanisation and to the return period of the events under consideration. Smaller events, particularly those of less than a few years return period, are proportionally more affected than those recurring, on average, less often (Hollis, 1975).

The provision of new drainage lines and the development of impermeable areas may be assisted in their enhancement of floodpeaks by more subtle effects of urban areas upon climatic regimes. Particularly in larger urban centres, rainfall totals are demonstrably higher than in the surrounding rural areas, with summer convectional storms notably affected. A summary of these effects is included by Douglas (1983).

Sediment loads from the catchment are also known to be increased manyfold by urbanisation, particularly during the surface disturbances of construction, but also after the cessation of this activity (Wolman and Schick, 1967; Walling and Gregory, 1970). With this complex shift in the controls on channel morphology, the channel might hypothetically be expected to respond by increasing its capacity, particularly once the high-sediment loads have been flushed through by successive flood events, but the length of time for complete adjustment is a matter of speculation. Leopold (1973) has demonstrated the persistence of in-channel sediments and reduced capacity in Watts Branch, Maryland, for at least 17 years after development. A schematic illustration of the channel adjustment process initiated by alterations to runoff, rainfall and sediment yields is illustrated in Figure 5.3.

ESTABLISHING THE EXTENT OF CHANNEL CHANGE

In Great Britain, the sources of historical documentation of changes in river channel morphology (Hooke and Kain, 1982) are often inappropriate by virtue of scale, coverage or degree of repetition for study of urban-induced change. Furthermore, the timescale of urbanisation usually precludes form or process monitoring during the relevant periods. Instead, reliance must be placed on indirect establishment of channel conditions prior to the urban development, and comparison with currently developed catchments. The controls on natural channel capacity being numerous and complex, only approximate estimates of rural capacity can be attempted, most simply by defining statistically the relationships between catchment characteristics and current channel dimensions in rural catchments. Any historical information such as early maps, surveys, or descriptions can then be placed in context as supporting evidence of pre-urban conditions.

The most fundamental catchment characteristic influencing channel capacity within a region is catchment area, acting through its influence on flood discharges, and it is this parameter that most previous research has adopted as a predictor of natural or rural channel capacity. When catchments of very variable

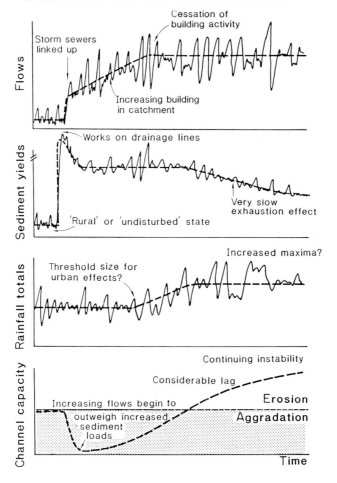

Figure 5.3 Hypothetical trends in the river system following urbanisation.

permeability are involved, total contributing channel length has been suggested as a better indicator (Petts, 1977), and channel slope is another possible direct influence on capacity (Hammer, 1972; Park, 1976) which may be incorporated into a statistical model. Whichever characteristic is adopted, a statistical relationship between the 'independent' variable and channel capacity may be established for rural upstream, or nearby, sites, and the relationship can then be extrapolated into the currently urbanised zone to allow comparison with channel capacity there. The inherent problems of statistical extrapolation, confidence limits, and natural channel variability, are compounded in many

Table 5.2 Parameters of regression model for rural channel capacity.

	Channel capacity (m^2)	Drainage area (km^2)	Geology index	Rainfall intensity (mm/h)	Drainage density (km km^2)	Winter rain acceptance potential	Basin compactness
Mean	−0.084	0.805	0.409	1.758	−0.0003	0.388	−0.063
Regression coefficients	(−3.298 intercept)	0.565	0.922	1.456	0.458	−0.304	0.557
Order of entry		1	4	3	2	5	6
Standard regression coefficients		0.824	0.464	0.446	0.362	−0.176	0.081

Correlation coefficient $r = 0.851$
Explained variance $r^2 = 73\%$
Standard error of estimate $= 0.163$
F-ratio $= 120.3$

Note: All figures are expressed in logarithmic units.

areas by the limited extent of 'natural' unaffected streams, or by problems of extension of the study area into different physiographic regions for which the established relationship may not be valid.

To circumvent some of these problems, the methodology adopted here for predicting rural capacity was similar to that utilised in the 'Flood Studies Report' (NERC, 1975) for predicting flood regimes from catchment characteristics, and drew partly on the same data set. Drainage area, contributing channel length, drainage density, basin compactness (Blair and Biss, 1967), channel slope, a geological substrate index, rainfall totals and intensity characteristics, winter rain acceptance potential and residual soil moisture deficit, were established for catchments feeding each of 382 rural channel cross-sections distributed throughout the study areas. A series of regression models were constructed, and the finally selected version incorporated only those 281 sites more than 50 m from artificial structures such as bridges, where no recent channel instability was evident in the field, and where catchment size exceeded 1 km^2. The headwater channels eliminated from the model at this state were very irregular in size, and extensively influenced by riparian vegetation. Their elimination did not materially alter the regression coefficients, but the level of statistical explanation was markedly improved. The model was thus developed from a wide range of geological, physiographic and climatic circumstances, and whilst it is clearly limited in its geomorphological explanatory powers, lacking detailed consideration of channel boundary materials for example, is adequate for the purposes of this study. Some details of the model are shown in Table 5.2.

The model suggests that drainage area is the most significant influence on capacity, with geology, drainage density, rainfall intensity characteristics, winter rain acceptance potential, and basin compactness also being significant. Several variables do not appear in the model because of collinearity problems, but this combination produces a statistically robust model, without major instability arising in the stepwise regression procedure, and with all variables logically signed. The statistical explanatory powers of the model are high, and the 'national' model is thus able to provide estimates of rural channel capacity for currently urbanised catchments. Estimates from the model were then compared with bivariate estimates using drainage area as the independent variable, and with any historical data available, thus establishing the best possible estimate upon which to base assessment of channel change.

An example of the use of this national model is illustrated in Figure 5.4. Cumbernauld New Town, Strathclyde, straddles the watershed of two catchments. Luggie Water drains westwards to the Clyde, collecting surface waters from a substantial area covered with thick boulder clays and peat, as well as urban drainage. Red Burn and its tributary Bog Stank drain the eastern part of the town, forming part of the Forth catchment. Cumbernauld was developed following designation in 1955, and has large areas of industrial park, with

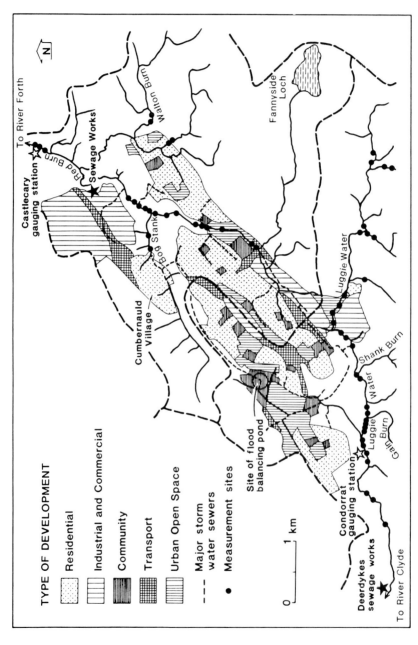

Figure 5.4 Cumbernauld New Town, Strathclyde. The urban development straddles the watersheds of Luggie Water and Red Burn.

densely developed housing and service areas. Utilising the national model for predicting channel capacity in the two catchments gives the estimates shown in Figure 5.5 as dotted lines. These may be compared with the bivariate model based on upstream rural sites in the locality, shown on the diagram with the appropriate standard error of forecast limits which are wide because of the relatively poor correlation between drainage area and channel capacity. The national model nevertheless supports the validity of locally based estimates, the two coinciding across the range illustrated. The national model may then be used as a basis for establishing an enlargement ratio for channel sections within the currently urbanised reaches of the two catchments. These are plotted in Figure 5.6, which suggests that many of the sections have eroded to between two and three times their previous size. The pattern of erosion in Luggie Water is a general increase in enlargement through the urban area, with a fall off downstream. A short reach apparently uninfluenced by enlargement (at 11–12 km downstream) was affected by canalisation in 1750, and erosion here is constrained. In Red Burn the enlargement appears to be much more irregular.

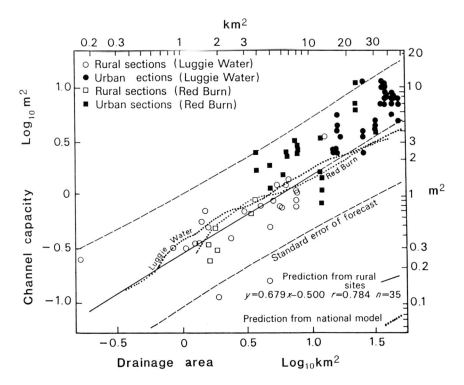

Figure 5.5 Comparison of locally derived (bivariate) and nationally derived (multivariate) estimates of rural channel capacity for Luggie Water and Red Burn.

68   Floods

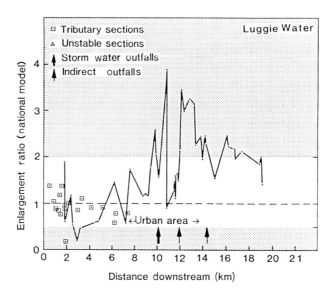

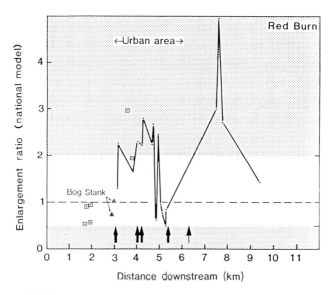

Figure 5.6 Enlargement ratios. Luggie Water and Red Burn, Cumbernauld.

PATTERNS OF EROSION AND SEDIMENTATION

The patterns of erosion found in Luggie Water and Red Burn are analogous to those revealed by field inspection of urbanised catchments across Britain. Many of the channels examined show characteristics which mimic those following major flood events. The general impression was of recent scour with undermining and widespread tree collapse, erosion around structures, undercutting of turf in grassed areas and 'hanging' non-urban tributary junctions. Locally there were instances of more massive erosion and bank collapse, usually associated with outfalls from surface water networks. Gully incision of steeply sloping tributary streams in receipt of substantial volumes of urban runoff was also present; in Cumbernauld up to 10 metres of vertical incision through glacial till was observed in steep gullies draining industrial areas. Some mainstream channels have developed incipient braiding, with substantial medial gravel bars deflecting the thalweg onto the banks, and apparently initiating changes in the channel pattern, but widespread changes to sinuosity were minimal. There were indications that the rate of change was greater than for comparable rural channels since raw, unvegetated banks were numerous, and that many of the urban channels had certain parameters still in a transient state (Neller, 1988).

Sedimentation

The pre-existing floodplain topography appears important in determining the nature of the response and some reaches experience overbank sedimentation with sand and silt rather than bank erosion. Abandoned channel loops (swales) and areas adjacent to channelised sections are common locations for overbank sedimentation, and it is not restricted to downstream reaches. Such sites may eventually achieve larger cross sections than their rural counterparts, although the relaxation time required to reach any new equilibrium state may be longer.

As has been established for the eastern US seaboard (Wolman, 1967; Hammer, 1972; Leopold, 1973) the process of adjustment to an urban regime appears to involve an initial period of in-channel sedimentation with debris derived from the catchment. In some areas, coarser elements become imbricated and partially preserved in floodplain sequences despite several decades of subsequent channel enlargement (the Huddersfield and Stevenage sites, for example), whilst in others the material is fine grained and unconsolidated, remaining as a blanket over pools and riffles until it is removed by a sequence of high flows (as in Leicester, for example). This may take several years to accomplish, depending on the hydrological pattern experienced in the particular catchment.

The origin of the different types of sediment within British channels remains unknown, although the coarser material is probably generated close to the channel network during the construction phase of urban development. Finer silts might continue to be produced within the urban area for long periods after construction had ceased, but the increased flows associated with storm sewerage and impermeable areas appear to be sufficient to remove them in later years. Sediment generated further from the channel will also tend to be trapped in road gully-pots. Natural channels in Britain rarely carry capacity loads of suspended solids and unless berms are stabilised by vegetation during a dry period, the material will be readily erodible.

Urban channel enlargement

Although enlargement appears to be the normal response of channels to the imposition of an urban regime, the extent of enlargement seems to vary with the nature of the channels. There is less visual difference between urban and rural channels, where relatively low runoff is associated with permeable catchments, than in those with impermeable basins. The incised, narrow, rectangular channels with stiff clay banks found in chalk or limestone catchments in southern England seem particularly resistant to change in their widths. However, the results of field inspection frequently give a misleading impression of the extent of channel change, especially where recent large floods have occurred, or where channels are naturally mobile, and it is important to obtain less subjective estimates.

Consideration of the mean enlargement ratios based on comparison of urban dimensions with modelled pre-existing rural dimensions for all the urbanised sites in individual catchments removes some of the variability associated with channel response at individual reaches and suggests the order of magnitude of change in each catchment, even though the channels may not have equilibrated fully with their new regimes. Care was taken to exclude from consideration any sites known to have been affected directly by engineering works. A total of 489 urban sites were surveyed and extensive data on the physical nature of urbanisation was collected for 430. At 59 sites, details of the storm drainage network layout were not available from the drainage authorities (usually district councils). Eleven further sites have tiny areas of urbanisation within their topographically defined catchment area, but do not receive surface drainage directly from them; these have been included in the analysis. The nature of the urban area and the relative influence of differing land use types and ages in British catchments is not presented here, but Hammer (1972) has provided some estimates for the United States. In Britain, the extent of stormwater sewerage seems the dominant criterion governing the extent of change, outweighing age, type and location of development in its effect.

The mean enlargement ratio (urban cross-sectional area/rural) is 1.607 (s.d. 0.950), with a range up to about 6. An isolated site in the River Tawd has a ratio of 8.75, and the whole data set is positively skewed. The mean rural enlargement ratio is 1.281 (s.d. 0.840), differing from unity because it is calculated as the

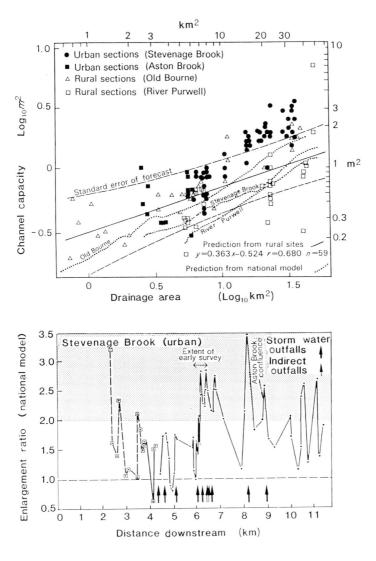

Figure 5.7 Channel capacity patterns and enlargement ratios, Stevenage Brook. Two adjacent rural catchments are shown for comparison.

mean of the unlogged enlargement ratios and includes the non-stable sites; deriving a mean from the logged figures, and then antilogging yields a figure of unity. The rural-urban difference of means is highly significant with $t = 5.355$ for a t-test with unequal population variances, or $K = 54.47$ for the non-parametric Kolmogorov–Smirnov test (d.f. = 2).

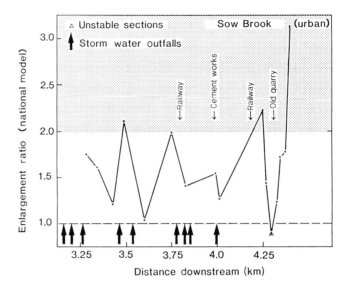

Figure 5.8 Enlargement ratios, Sow Brook, Rugby.

It is possible that the national enlargement model is producing some systematic underestimate of preexisting channel dimensions and thus yielding spuriously large enlargement ratios for the urban sites, but generally the agreement between the national prediction model and historical data sources is good, suggesting that the enlargement ratios are of the correct order of magnitude. For isolated catchments, however, the prediction of rural channel capacity is very poor. Pin Brook, for example, appears to have an overly large capacity in both rural and urban reaches. This is undoubtedly the result of a flood estimated to have a recurrence interval in excess of 30 years which occurred in July 1972, four years before the field study was undertaken. Hence there is no detectable 'urban' influence on the channel. The sites in the Red River catchment near Camborne, severely affected by sediment pollution, conversely have exceptionally small capacities. These and similar channels were removed from subsequent analysis.

Examples of the patterns of channel capacity in urban (and adjacent rural) catchments are shown in Figures 5.7-5.9, from which the great variability in channel response is clear. In Stevenage Brook, below the New Town, the average enlargement ratio based on the national model is slightly under 2.0, in the same range as suggested by comparison with adjacent rural catchments (and covered in my earlier paper—Knight, 1979). Enlargement appears rather more marked in downstream reaches, where stream power is higher, than in those where the major urban inputs are made. In Sow Brook, a tributary of the Avon in the west of Rugby, the extent of enlargement appears to be similarly modest, despite the

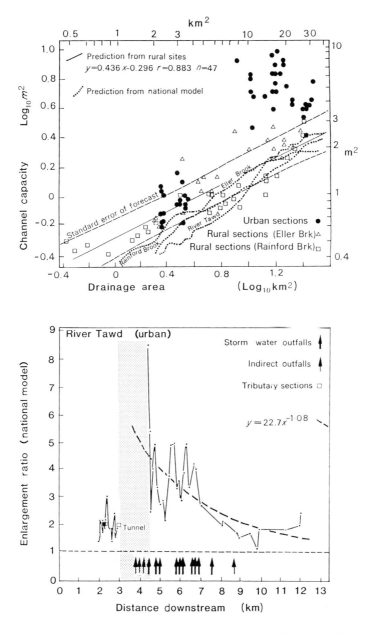

Figure 5.9 Channel capacity patterns and enlargement ratios, River Tawd, Skelmersdale, Lancashire. Two adjacent rural catchments are shown for comparison.

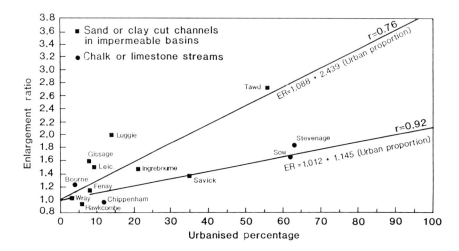

Figure 5.10 Mean enlargement ratio as a function of mean urbanised percentage of the catchment.

relatively large urban percentage (Figure 5.8). Both Stevenage Brook and Sow Brook are clay-bank channels with limited mobile bed sediments, flowing in permeable catchments. By contrast, the River Tawd which drains almost all of Skelmersdale New Town, Lancashire, exhibits a very marked enlargement (Figure 5.9) which can be confirmed in some reaches by comparison with an engineering survey carried out prior to development (Knight, 1979).

Relationships with the extent of urbanisation

Figure 5.10 shows the general relationship between the mean enlargement ratios and the mean proportion of the catchments with urban land uses, including open space (*Department of Environment Developed Areas Survey*, 1975). Catchments with little urbanisation (Wray Brook, River Bourne, Fenay Beck) show low average enlargement, with ratios around unity, whereas catchments with substantial urban areas (Ingrebourne, Savick Brook, Leicester streams, River Tawd) show consistently greater enlargement. There is no indication of any threshold in the response to urbanisation as was suggested by Morisawa and LaFlure (1979) for the US but catchments in chalk or limestones tend to show lesser effects, despite having substantial proportions of urban use in some cases (Stevenage Brook, Sow Brook, Chippenham). Regression relationships for the two groups are significant (at 1%) and demonstrate almost perfect proportionality between enlargement and urbanisation; complete absence of bias in the equation specification should produce zero intercepts. For the 'impermeable' catchment group,

E.R. = 1.088 + 2.439 (urbanised proportion) (r = 0.76)

and for the 'permeable' group,

E.R. = 1.012 + 1.145 (urbanised proportion) (r = 0.92).

It is important to note that the distinction into groups is not so much a reflection of the catchment characteristics (permeability, for example) as a reflection of the power of the flood discharges generated in relation to the resistance of the channel boundary. Permeability has been fully reflected in the calculation of enlargement ratio and so the permeable–impermeable distinction is not distinguishable in patterns of residuals from the national model fit to rural channel sections. Ferguson (1981) has noted that related planform inactivity in such feeble channels may also be the result of discharges currently reduced well below those encountered in the early Holocene. For 100% urbanisation of the catchment, the mean enlargement ratio will be about 3.53 in a high-power, 'impermeable' channel with relatively low bank resistance, and 2.16 in a low-power, 'permeable' area channel with tough clay banks. A similar analysis using the mean sewered proportion of the catchment shows the distinction even more clearly (Figure 5.11). For the 'impermeable' group,

E.R. = 1.182 + 3.638 (sewered proportion) (r = 0.86)

and for the 'permeable' group,

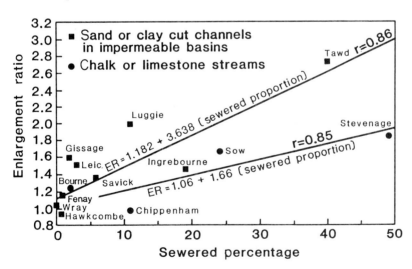

Figure 5.11 Mean enlargement ratio as a function of mean sewered percentage of the catchment.

E.R. = 1.060 + 1.660 (sewered proportion) ($r = 0.85$).

Both relationships are significant despite the small sample in the latter group. Complete sewering of the catchments would produce enlargement ratios of 4.82 and 2.72 respectively. Because of collinearity between urbanisation and sewering it is not possible to distinguish the relative effects of each factor, and each equation assumes either an average amount of sewering or a representative mix of land uses within the urban area, in addition to the independent variable variation. Naturally, further case studies would strengthen this hypothesis.

Skelmersdale and Cumbernauld New Towns have slightly more enlargement than their urban characteristics might suggest; this could be a reflection of their relatively coherent surface water drainage planning, compared with other urban areas. The effect might be underestimated because of their relatively recent development. The Bourne below Tidworth Garrison also shows a relatively sizeable effect when its urban characteristics are taken into consideration; coherent and intensive surface water drainage is again found here. By contrast, Chippenham and the London suburbs draining to the Ingrebourne are not efficiently sewered, piecemeal development having been added into the upper ends of existing networks. This has frequently led to blockages, surcharging and floodpeaks less enhanced than might otherwise be anticipated. It is important to note that the effects of variations in urban area or extent of sewering appear to mask the effects of variations in the length of time available for adjustment to occur. This implies that the majority of channel change may be accomplished relatively rapidly.

The results of this analysis are similar in magnitude to some studies of urbanised channels, but differ markedly from others. Not all the published studies give any indication of the urbanised or sewered fraction of the catchments involved, making comparison difficult. The accuracy of determination of enlargement ratios is in some studies questionable, the channels under consideration often very small, and there are no published studies dealing explicitly with channels in chalk catchments. Generally, there is agreement with the orders of magnitude of change suggested by Hollis and Luckett (1976) for West Sussex streams, by Mosly (1975) for the Bollin-Dene and for a range of North American examples (Hammer, 1972; Robinson, 1976; Morisawa and LaFlure, 1979). Studies by Gregory (1976, 1977), Gregory and Park (1976), and Park (1977) have tended to suggest much more extreme responses to small areas of urbanisation than the present study.

DISCUSSION

When urbanisation alters catchment response to rainfall, some of the increase of peakflow or channel modifying discharge is taken up by velocity adjustments by virtue of changes in channel geometry and roughness. This tends to minimise increase in width and depth.

Cross-sectional changes can also be complex with low-flow berms constraining baseflow but increasing sediment-carrying capacity, and being swept away at higher flows. Richards and Wood (1977), for example, suggest that for a doubling in mean annual flood (which is assumed to be the channel-forming discharge), width may increase by 20 to 50%, depending on the pre-existing channel equilibrium and hydraulic geometry exponents. Usually the change in channel dimensions would be slightly less than the proportional increase in channel-forming (bankfull) discharges.

The preliminary results of the present study reinforce this theoretical picture. In the case of the 'impermeable' channel group, enlargement is of similar order of magnitude to the increases in frequent peakflows anticipated by Leopold (1968) in the United States. For the 2.33-year flood (annual series) in a 2.58 km^2 basin, 25% impervious and 40% sewered, the peakflow multiplication factor would be about 2.2. Hammer's (1973) hydrological results are similar. For Britain, Packman (1979) estimates the peakflow multiplication factor to be roughly equal to $(1 + \text{urban proportion})^{1.8}$, yielding a factor of 2.07 for 50% urbanisation. Consequently, it may be assumed as a first approximation that the effects of channel storages, downstream dissipation of erosional effects, and complex erosional and depositional responses feeding up and down the channel are minimal; channel enlargement is usually a direct and proportional response to change in flood peaks. There remains, however, the problem of the apparently unresponsive nature of the 'chalk' channel group.

Flood frequency and urbanisation

An explanation of the differential extent of erosion in the 'impermeable' and the 'permeable' group may lie in the return periods of flood events to which channels are adjusted, and the extent to which these are modified by urbanisation. Hollis's (1975) analysis of the effects of urbanisation on floods of different return periods reveals that short-term events are much more affected than longer-term ones, and that the curve relating peakflow enhancement to recurrence interval falls particularly steeply in the range one to two years (annual series). For the 1-year event, the enhancement is 10 times for 20% paving (40% urbanisation, Hollis's criteria), whereas the 2-year event is only doubled or trebled. The 2.33-year event or mean annual flood would be about 2.5 times its former magnitude, as found at Stevenage and Skelmersdale, both of which have similar amounts of urban area (Knight, 1979). Five percent paving of the basin only affects floods of recurrence intervals shorter than one year. Although Packman's (1979) analysis failed to confirm this for British rivers alone, and his published growth curves suggest a similar order of enhancement for both the 2- and the 5-year flood with 50% urbanisation, this may be the result of bias in the data set (London is overrepresented) and short records with only slight urbanisation. Generally, short return period events are more seriously affected by urbanisation than more extreme events, presumably because a catchment in its natural state may respond as if impermeable when large areas are saturated by throughflow or interflow.

Bankfull channel capacity and flood frequency

The maintenance of bankfull channel capacity and its relationships with various parameters of flow have been the subject of much research, both in Britain and more widely, and it is well summarised by Richards (1982). Ignoring the separate adjustments of width and depth which seem more strongly associated with sediment parameters, and meander wavelength which may be influenced by rock type, most researchers suggest that natural bankfull capacity is adjusted to flows which recur in the range once every few months to once every few years. In reality a range of flows are involved, but these are represented by the single flow. This flow is not necessarily equivalent to the flow performing most erosion or moving most bed material (the so-called 'dominant' discharge). Neither is it related to the morphology of extensive 'fossil' terraces which are only flooded very exceptionally.

The earlier geomorphological studies suggested recurrence intervals of 1–2 years (annual series) to be virtually universal for bankfull flow, a figure subsequently refined to a mean of 1.58 years. This latter represents the most probable annual flood on Gumbel Type I distributions, and corresponds roughly to the one-year event on a partial series. It was subsequently suggested that steeper channels had higher bankfull recurrence intervals, and that the geology of the underlying catchment might have some influence through the nature of the flow regime. For flashy catchments on boulder clay, bankfull frequencies of 1.8 years were suggested, whilst on chalk overtopping occurs only in the seven-year event (annual series). Catchments on mixed geologies occupy intermediate positions (Harvey, 1969).

Many authors have suggested further extensions to the possible range of values for the recurrence interval of bankfull discharge, from a minimum of as little as a couple of months to as much as thirty years or more (Williams, 1978). Most studies either state or imply that the relationships are independent of the size of the catchment, although some suggest greater frequencies of overbank flooding in downstream sections of a river, in comparison with headwaters, whilst others hypothesise downstream adjustment to greater return period events because of the increasing duration of bankfull flows caused by floodpeak attenuation. There is no real indication of a size-related effect in the British catchments studied here, nor of any association with channel slope, although the permeable catchments also tend to be flatter. The recurrence interval of bankfull discharge thus appears to be some function of the flashiness of the hydrological regime, and thus indirectly of the basin geology.

Average enlargement and flood frequency

Hydrological data for the studied catchments is limited to relatively short non-stationary records in some cases, and is completely absent for others, so that inference can only be based on a few examples. Considering sites adjacent to gauging stations, the Stevenage channel appeared to be adjusted to a flow with a recurrence interval of about 2.8 years (partial series) both before and after urbanisation (Knight, 1979). The Bourne, monitored at Laverstock Mill in its lower perennial reach, has a bankfull flood frequency in excess of two years, and observation suggests a much longer interval further upstream. Sow Brook, flowing largely on limestones, also has a low bankful flood frequency. Of the less permeable catchments, Luggie Water (at Condorrat) experienced overbankfull events at an average interval of 4 months, in 8 years of record. The Tawd at Skelmersdale (Knight, 1979) is also adjusted to a much shorter return period event (8 months before urbanisation, 6 months afterwards, assuming no major changes in roughness). Savick Brook near Preston flooded several times per year according to diaries of local residents 40 years ago and the Ingrebourne (at Gaynes Park) has a bankfull flood frequency of about 1.33 years.

Assuming that most sites in the surveyed streams are approaching a new equilibrium, and mindful of the dangers of circularity in establishing bankfull flows from channel dimensions, the analysis suggests that average enlargement within the channel is strongly influenced by the flood frequency to which channels are adjusted. The 'impermeable catchment' group show roughly twice as much enlargement as the 'permeable' group with equivalent urban area, because they are adjusted to floods with shorter return periods which are preferentially influenced by urbanisation, as Hollis (1975) has demonstrated. Furthermore, within the 'impermeable' group there are indications that those channels with extremely short bankfull return periods (Luggie Water, Tawd) have experienced greater relative amounts of erosion than their less flashy counterparts.

CONCLUSION

An explanation of the extent of urban-induced channel enlargement in terms of the major characteristics of the developed area, and the nature of flood frequency relationships in the catchment, can of necessity only be a preliminary statement. It ignores completely many other urban-induced effects, including alterations in the timing of flood peaks, hydrograph shapes and their sediment concentrations and particle size characteristics. Urbanisation on chalk catchments, for example, may produce double-peaked hydrographs with urban-derived water passing from the catchment before the arrival of the main floodwave. In reality too, the response of the channel will be complex with feedback up and down the network perhaps creating different cycles of change at one location in comparison with another. Sediment generated by erosion at upstream locations must pass through lower sites before being lost from the catchment. Temporary storage in the lower channel may alter channel characteristics both locally and upstream, by reducing the water surface gradient and the flow velocity. Moreover, adjustment in the planform may take place more slowly than adjustment in the cross-section, and could initiate a subsequent further change in these latter parameters.

The overall timescales required for cross-sectional change will also vary depending on the actual sequence of flood events and may reflect the ease with which channel banks are saturated by exposure to high, but rapidly oscillating, short-duration flows. Vegetative constraints on erosion rates will also be important at some sites. Nevertheless, the fact that mean enlargement ratios are generally explicable in terms of simple urban parameters such as the urbanised or sewered percentages, and changes in flood frequency, suggests that the relaxation times involved for most of the change to be accomplished are relatively short, of the order of a few decades, for many British rivers.

REFERENCES

Blair, D. J., and Biss, T. H. (1967). The measurement of shape in geography: an appraisal of the methods and techniques, *Bull. Quant. Data for Geographers,* **11**, University of Nottingham.

Brookes, A. (1985). River channelization, *Progress in Physical Geography*, 9(1), 44–73.

Crippen, J. R., and Waananen, A. O. (1969). *Hydrologic Effects of Suburban Development near Palo Alto, California*, U.S. Geological Survey Open File Report.

Douglas, I. (1983). *The Urban Environment*, Arnold.

Ferguson, R. I. (1981). Channel form and channel changes, in J. Lewin, (ed.), *British Rivers*, George Allen and Unwin, pp. 90–125.

Fox, H. L. (1976). The urbanising river: a case study in the Maryland Piedmont, in D. R. Coates, (ed.), *Geomorphology and Engineering*, George Allen and Unwin, pp. 245–71.

Graf, W. L. (1975). The impact of suburbanisation on fluvial geomorphology, *Water Resources Research*, **11**, 690–2.

Gregory, K. J. (1976). Changing drainage basins, *Geog. J.,* **142**, 237–47.

Gregory, K. J. (1977). Channel and network metamorphosis in Northern New South Wales, in K. J. Gregory, (ed.), *River Channel Changes*, Wiley, pp. 389–410.

Gregory, K. J., and Park, C. C. (1976). Stream channel morphology in north west Yorkshire, *Rev. de Geom. Dyn.,* **25**, 63–72.

Hammer, T. R. (1972). Stream channel enlargement due to urbanisation, *Water Resources Research,* **8**, 1530–40.

Hammer, T. R. (1973). Impact of urbanisation on peak streamflow, *Regional Science Research Institute Discussion Paper Series*, **73**, 77pp.

Harvey, A. M. (1969). Channel capacity and the adjustment of streams to hydrologic regime, *J. Hydrol.,* **8**, 82–98.

Helliwell, P. R., and Kidd, C. H. R. (1979). Storm flows in suburbia: the Southampton small catchment experiment, in G. E. Hollis, (ed.), *Man's Impact on the Hydrological Cycle in the United Kingdom*, Geobooks, pp. 173–80.

Hollis, G. E. (1974). The effect of urbanisation on floods in the Canon's Brook, Harlow,

Essex, in K. J. Gregory and D. E. Walling, (eds), *Fluvial Processes in Instrumented Catchments*, Inst. Brit. Geog. Spec. Publ. 6, pp. 123-39.

Hollis, G. E. (1975). The effect of urbanisation on floods of different recurrence intervals, *Water Resources Research,* **11**, 431-5.

Hollis, G. E., and Luckett, J. K. (1976). The response of natural stream channels to urbanisation: two case studies from south east England, *J. Hydrol.,* **30**, 351-63.

Hooke, J. M., and Kain, R. J. P. (1982). *Historical change in the physical environment: a guide to sources and techniques*, Butterworths.

Knight, C. R. (1979). Urbanisation and natural stream channel morphology: the case of two English new towns, in G. E. Hollis, (ed.), *Man's Impact on the Hydrological Cycle in the United Kingdom*, Geobooks, pp. 181-98.

Leopold, L. B. (1968). *Hydrology for Urban Land Planning-a Guidebook on the Hydrologic Effects of Urban Land Use*, U.S. Geological Survey Circular 554, pp. 1-18.

Leopold, L. B. (1973). River channel change with time: an example, *Bull. Geol. Soc. of America,* **84**, 845-60.

Morisawa, M. E., and LaFlure, E. (1979). Hydraulic geometry, stream equilibrium and urbanisation, in D. D. Rhodes and G. P. Williams, (eds), *Adjustments of the Fluvial System*, Allen and Unwin, pp. 333-50.

Mosly, M. P. (1975). Channel changes on the River Bollin, Cheshire, 1872-1973, *East Midland Geographer,* **6**, 185-99.

Nanson, G. C., and Young, R. W. (1981). Downstream reduction of rural channel size with contrasting urban effects in small coastal streams of southeastern Australia, *J. Hydrol.,* **52**, 239-55.

Neller, R. J. (1988). A comparison of channel erosion in small urban and rural catchments, Armidale, New South Wales, *Earth Surface Processes and Landforms,* **13**, 1-7.

NERC, 1975, Flood Studies Report, 5 volumes Natural Environment Research Council, Wallingford.

Packman, J. C. (1979). The effect of urbanization on flood magnitude and frequency, in G. E. Hollis, (ed.), *Man's impact on the Hydrological Cycle in the United Kingdom*, Geobooks, pp. 153-172.

Park, C. C. (1976). The relationship of slope and stream channel form in the River Dart, Devon, *J. Hydrol.,* **29**, 139-47.

Park, C. C. (1977). Man-induced changes in stream channel capacity, in K. J. Gregory, (ed.), *River Channel Changes*, Wiley, pp. 121-44.

Petts, G. E. (1977). Channel response to flow regulation: the case of the River Derwent, Derbyshire, in K. J. Gregory, (ed.), *River Channel Changes*, Wiley, pp. 145-64.

Richards, K. S. (1982). *Rivers: Form and Process in Alluvial Channels*, Methuen.

Richards, K. S., and Greenhalgh, C. (1984). River channel change: problems of interpretation illustrated by the River Derwent, North Norkshire, *Earth Surface Processes and Landforms,* **9**, 175-80.

Richards, K. S., and Wood, R. (1977). Urbanisation, water redistribution and their effect on channel processes, in K. J. Gregory, (ed.), *River Channel Changes*, Wiley, pp. 369-88.

Robinson, A. M. (1976). The effects of urbanisation on stream channel morphology, in *National Symposium on Urban Hydrology, Hydraulics, and Sediment Control*, University of Kentucky, Lexington.

Thornes, J. B. (1977). Hydraulic geometry and channel change, in K. J. Gregory, *River Channel Changes*, Wiley, pp. 91-100.

UNESCO (1979). *Impact of urbanisation and industrialisation on water resources planning and management*, Report of the UNESCO/IHP workshop at Zandvoort, The Netherlands, 1977, UNESCO.

Walling, D. E., and Gregory, K. J. (1970). The measurement of the effects of building construction on drainage basin dynamics, *J. Hydrol.*, **11**, 129–44.

Whipple, W., and DiLouie, J. (1981). Coping with increased stream erosion in urbanising areas, *Water Resources Research*, **17**(5), 1561–4.

Williams, G. P. (1978). Bankfull discharge of rivers, *Water Resources Research*, **14**(6), 1141–54.

Wolman, M. G. (1967). A cycle of sedimentation and erosion in urban river channels, *Geografiska Annaler*, **49**, Series A, 385–95.

Wolman, M. G., and Schick, A. P. (1967). Effects of construction on fluvial sediment, urban and suburban areas of Maryland, *Water Resources Research*, **3**, 451–64.

6 Hydraulics of Flood Channels

DR DONALD W. KNIGHT
Civil Engineering Department, University of Birmingham, Birmingham

INTRODUCTION

Natural rivers have always been at the centre of civilisation. They provide a convenient access route to the sea, water for irrigation, a potential supply of drinking water, a possible source of power and, of course, a certain amount of aesthetic charm. They also perform a primary function in land drainage and a possible secondary role in waste disposal. For these reasons communities have often been built adjacent to rivers or on flood plains, where the soil is naturally fertile due to previous inundations. However, the benefits of such a geographical location have to be weighed against the possible hazards, one of which is certainly the risk of flooding. The cost/benefit ratio is therefore an important economic indicator in the design of most flood alleviation works. For this reason the hydraulics of flood channels is a topic which is taken seriously by Civil Engineers as illustrated by recent conferences (BHRA, 1983; BHRA, 1985; Smith, 1984). Urban drainage and channel improvement schemes have also highlighted the need for good hydraulic engineering practice in man-made systems.

From a geomorphological viewpoint, flows in channels and over land surfaces are one of the dominant factors in producing change. If these changes are to be quantified then some knowledge of hydraulic engineering is required. This brief review of certain hydraulic aspects of flood channels is therefore aimed at acquainting the reader with some of the basic ideas in open channel flow. Three topics have been selected for comment in this paper and are as follows:

(1) stage discharge curves;
(2) in-bank flows;
(3) out-of-bank flows.

A comprehensive list of references is provided for the reader who wishes to learn more about the general principles of open channel flow and flood hydraulics in particular.

STAGE-DISCHARGE CURVES

Introduction

The relationship between stage and discharge (h and Q) for a river or open channel is determined by correlating measurements of discharge with corresponding observations of stage. Once determined, this relationship is of great value since it may be used to infer discharges at particular stages or to estimate levels at extreme discharges under flood flow conditions. The relationship will be unique, provided that the channel is stable and the flow steady. A stable channel is defined as one in which the physical form, control characteristics and frictional properties of the bed and sides remain constant with respect to time. Under these conditions the stage–discharge relationship may be expressed (Herschy, 1978, ISO, 1982) in the general form

$$Q = C(h + h_o)^n \tag{6.1}$$

where Q = discharge (m^3/s)

h = stage (m)

h_o = zero correction (m)

C, n = parameters to be determined by observation.

It should be noted that h_o is a correction factor which needs to be determined if the zero of the gauge does not coincide with zero discharge. For the special case of $h_o = 0$, the stage h is equivalent to the depth of flow in the channel. A typical stage discharge curve with the equation

$$Q = 39.479 \, (h - 0.115)^{1.530} \tag{6.2}$$

is shown in Figure 6.1(a). Thus for $h = 1.0$ and 3.0 m, $Q = 32.748$ and 199.705 m^3/s respectively.

Normal depth flow

The stage–discharge relationship for simple prismatic channels may be readily obtained by applying an appropriate uniform flow resistance formula such as

Chezy	$Q = CAR^{1/2} S_f^{1/2}$	(6.3)
Manning	$Q = AR^{2/3} S_f^{1/2} / n$	(6.4)
Darcy-Weisbach	$Q = (8g/f)^{1/2} AR^{1/2} S_f^{1/2}$	(6.5)

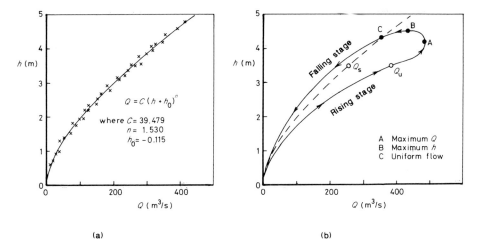

Figure 6.1 Stage-discharge curves: a) steady flow; b) unsteady flow.

where A = area of cross-section, R = hydraulic radius (= A/P), P = wetted perimeter, S_f = energy slope (= bed slope, S_o, under uniform flow conditions) and C, n and f are resistance coefficients. It is known that f varies with Reynolds number $(4UR/v)$ and relative roughness $(k_s/4R)$ according to the Colebrook–White equation (1939)

$$\frac{1}{\sqrt{f}} = -2.0 \log \left[\frac{k_s}{14.84R} + \frac{2.51}{(4uR/v)\sqrt{f}} \right] \quad (6.6)$$

where k_s = Nikuradse's equivalent sand roughness size and v = kinematic viscosity. Equation (6.6) may be combined with equation (6.5) to give the Darcy–Weisbach equation in terms of the mean flow parameters:

$$Q = -A\sqrt{(32gRS_f)} \log \left[\frac{k_s}{14.84R} + \frac{1.255v}{R\sqrt{(32gRS_f)}} \right] \quad (6.7)$$

In applying equations (6.3)–(6.7) it should be noted that the hydraulic parameters A, P and R are related to the depth of flow h by ancillary equations which depend upon the shape of the cross-section. Thus for a trapezoidal channel of base width, b, side slopes $1:s$ (vertical:horizontal) and top width T, the ancillary equations would be

$$A = (b + sh)h$$
$$P = b + 2h\sqrt{(1 + s^2)}$$

$$R = A/P$$

$$T = b + 2sh \tag{6.8}$$

Inserting these into equation (6.4) by way of illustration, collecting terms which contain geometric parameters on the right-hand side, and dividing throughout by $b^{8/3}$ in order to make the formulation dimensionless, gives the following equation:

$$\frac{Qn}{b^{8/3} S_f^{1/2}} = \frac{(b + sh)^{5/3} h^{5/3}}{b^{8/3} \left[b + 2h \sqrt{(1 + s^2)} \right]^{2/3}}. \tag{6.9}$$

Solution of this equation gives the theoretical h versus Q relationship provided the Manning roughness coefficient n is known. If the equation is to be solved for h, given the value of Q, then an iterative procedure or design chart is required: (See for example Chow, 1959; French, 1986; Henderson, 1966 or Subramanya, 1982). If the fixed point iteration procedure is adopted, then equation (6.9) may be written in the following form, which converges rapidly after a few trials:

$$\frac{h_{j+1}}{b} = \left[\frac{Qn}{b^{8/3} S_f^{1/2}} \right]^{3/5} \frac{b^{3/5} \left[b + 2h_j \sqrt{(1 + s^2)} \right]^{2/5}}{(b + sh_j)} \tag{6.10}$$

where h_{j+1} = new value of depth
h_j = old value of depth.

An initial guess of $h_j = 0$ will always be suitable for this case since the first h_{j+1} value will then correspond to the wide rectangular channel case. However, for the special case of uniform flow in a triangular channel ($b = 0, s > 0$) this iterative procedure breaks down and the exact analytical solution should be used. The value of h finally obtained from this procedure is the so-called 'normal depth', h_n, which is the depth of flow which will occur in a channel when the gravitational force resolved down the channel slope (= $\rho g A S_f$ per unit length) just balances the frictional resistance around the wetted perimeter (= $\tau_o P$ per unit length). Under these circumstances the mean boundary shear stress acting on the channel wetted perimeter, τ_o, is then given by

$$\tau_o = \rho g (A/P) S_f = \rho g R S_f \tag{6.11}$$

and the corresponding mean shear velocity, $u_* \ (= \sqrt{(\tau_o/\rho)})$ may be easily determined.

Although this procedure is widely used in hydraulic engineering to calculate normal depth flow in canals, culverts, sewers running part-full and other prismatic channels, its application to natural rivers is not as straightforward as it might appear since the procedure assumes that

(1) the roughness coefficient is known;
(2) the roughness coefficient does not vary with stage;
(3) the cross-section does not change markedly with stage;
(4) the energy slope is known;
(5) the flow is steady and uniform.

Unless these conditions are satisfied, or that reasonable assumptions can be made about k_s, f, n or C, it is generally safer in practice to determine the h versus Q relationship by direct measurement as shown for example in equation (6.2). However, when this is not possible and the normal depth calculation procedure or equations (6.3) to (6.4) are used, then it is important to appreciate certain technical details concerning open-channel flow resistance.

Open-channel flow resistance

Channels may be broadly classified according to the nature of their boundaries, which might be rigid, loose or flexible:

(1) *Rigid boundary* (e.g. concrete channel, lined canal, non erodible earth). Despite a fixed geometry, the mean resistance coefficient for a reach is influenced by the shape of the cross-section, the non-uniformity of the textural roughness around the wetted perimeter, and the form drag arising from three-dimensional effects. These are illustrated individually and schematically in Figures 6.2(a)–(c). It should be remembered that equation (6.6) is only strictly valid for flows in circular pipes, and that the use of the hydraulic radius, R, is not always appropriate, particularly for a cross-sectional shape like that in Figure 6.2(a). As the water flows on to the berm, the wetted perimeter increases rapidly without a significant change in cross-sectional area. The influence of shape on the Darcy–Weisbach resistance coefficient is well documented (Jones, 1976; Lai, 1987; Rehme, 1973; Schlichting, 1979). Figure 6.2(b) indicates that the textural roughness may differ between the main channel and the flood plains. Where the roughness or shape changes are not too rapid, then Pavlovskij's method (Chow, 1959) may be adopted to calculate a composite mean value of roughness. In cases where the roughness of the sides and bed of a channel differ significantly or the river is ice covered, more refined methods are required (Knight, 1981; Larsen, 1973; Naot, 1984). Figure 6.2(c) is a reminder that a single-valued reach mean resistance coefficient often includes a component of form drag due to three-dimensional effects arising from flow around bends, abrupt transitions and overbank flow. A single-valued resistance coefficient is thus a 'lumped

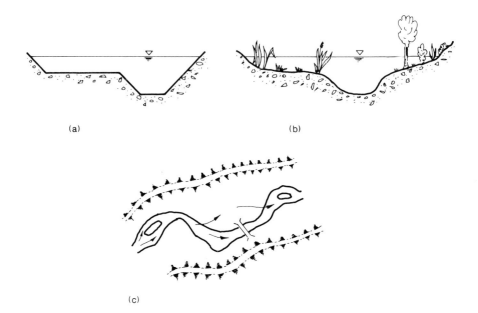

Figure 6.2 Rigid boundary resistance effects: a) non-uniform shape; b) non-uniform roughness; c) non-uniform plan form geometry.

coefficient', and includes a component of textural, form and plan geometry resistance.

(2) *Loose boundary* (e.g. erodible natural rivers with silt, sand, gravels and boulders). When water flows over a mobile boundary, and the threshold boundary shear stress is exceeded, the fluid interacts with the bed and two-phase flow occurs. For sand bed channels the bed no longer remains plane but develops through a well-defined series of bed forms which introduce additional form drag. This is illustrated in Figure 6.3. The total resistance to flow is then composed of two components, a textural component, f', and a form drag component, f''. The total resistance $f(=f'+f'')$ varies with flow rate in a complex manner (ASCE, 1963; Garde and Ranga Raju, 1977; Hey et al., 1982; Shen, 1979; Vanoni, 1975). This is illustrated in a simplistic way in Figure 6.3(b), where the component f'' is seen to increase rapidly in lower regime flow once ripples and dunes form, then decrease as transitional or plane bed conditions occur, before increasing again in upper regime flow when antidunes and other bed forms may occur. The influence of changing bed forms may have a significant effect upon stage discharge curves, as illustrated by the Padma River in Bangladesh (Shen, 1979). In addition to the changes in roughness, a loose boundary channel is by definition unstable, and patterns of erosion, deposition,

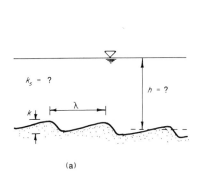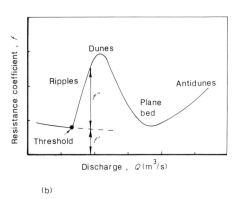

Figure 6.3 Loose boundary resistance effects: a) bed forms; b) textural and form drag components.

plan form geometry change, and changing hydraulic control all may influence the stage discharge curve.

(3) *Flexible boundary* (e.g. grass, vegetal growth on flood plains). Weed growth significantly affects the conveyance capacity of rivers. There are strong seasonal variations arising from the natural processes of growth and decay and also possibly from weed cutting programmes undertaken by drainage authorities. In times of flood flow, grasses or other tall vegetation may be flattened temporarily, thus reducing the resistance considerably. This is illustrated schematically in Figure 6.4. Some guidelines concerning flow resistance of grasses are available. (Kouwen and Li, 1980; Ree and Palmer, 1949). It should also be remembered that loose floating debris, such as bushes and trees, often causes blockage of drainage channels and creates localised flooding.

Unsteady flow

By definition flood flows are unsteady. Flood hydrographs usually exhibit a rapid rise followed by a slow recession. The translation, attenuation and deformation of the hydrograph is of considerable importance when modelling flood flows down rivers (Price, 1985). The unsteadiness of the flow is also of importance in flow gauging since observations of stage correlate with different discharges depending upon whether the river is rising or falling. This is best understood in relation to the equations

$$Q_s = \frac{AR^{2/3}S_f^{1/2}}{n} = \frac{AR^{2/3}S_o^{1/2}}{n} \tag{6.12}$$

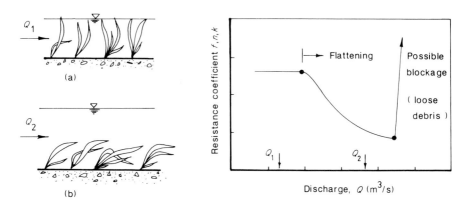

Figure 6.4 Flexible boundary resistance effects: a) low stage; b) high stage.

$$Q_u = \frac{AR^{2/3}}{n}\left[S_o - \frac{\partial h}{\partial x} - \frac{u}{g}\frac{\partial u}{\partial x} - \frac{1}{g}\frac{\partial u}{\partial t}\right]^{1/2} \quad (6.13)$$

$$\cong \frac{AR^{2/3}}{n}\left[S_o - \frac{\partial h}{\partial x}\right]^{1/2} \quad (6.14)$$

where Q_s = steady flow, Q_u = unsteady flow, S_o = bed slope, S_f = friction or energy slope, u = section mean velocity (= Q/A), h = depth of flow, x = longitudinal coordinate and t = time. Equation (6.12) shows that for uniform flow $S_f = S_o$, whereas equation (6.13) shows that for non-uniform unsteady flow S_f has at least four components based on the simple one-dimensional momentum equation. Since in many natural rivers the convective and local acceleration terms are small, equation (6.13) is often reduced to the form shown by equation (6.14). Moreover, since

$$\frac{\partial h}{\partial x} = -\frac{1}{c}\frac{\partial h}{\partial t} + \frac{dh}{dx}$$

$$\frac{Q_u}{Q_s} = \left[1 + \frac{1}{cS_o}\frac{\partial h}{\partial t} - \frac{1}{S_o}\frac{dh}{dx}\right]^{1/2}$$

$$\therefore \frac{Q_u}{Q_s} \cong \left[1 + \frac{1}{cS_o}\frac{\partial h}{\partial t}\right]^{1/2}. \quad (6.15)$$

Equation (6.15) thus shows that for given values of kinematic wave speed, c and bed slope, S_o, $Q_u > Q_s$ depending upon the sign of $\partial h/\partial t$. For a rising stage

Hydraulics of Flood Channels 91

$\partial h/\partial t$ is positive, making $Q_u > Q_s$, whereas for a falling stage $\partial h/\partial t$ is negative and $Q_u < Q_s$. This produces the well-known looped rating curve as illustrated by Figure 6.1(b). During the passage of a flood event an observer would therefore see firstly the maximum discharge (point A), followed by the maximum stage (point B) and then finally the equivalent uniform flow value of stage (point C). The looped rating curve thus implies that flow gauging reecords should always be corrected for unsteady flow effects. It also has implications for estimating discharge values from geological stage records. Assuming a range of parameter values of $0.3 < u < 3.0$ m/s, $0.5 < c < 5.0$ m/s and $10^{-3} < S_o < 10^{-5}$, then taking nominal values of $u = 0.6$ m/s, $c = 1.0$ m/s, $S_o = 1.0 \times 10^{-3}$ and $\partial h/\partial t = 1.0$ m in 1 hour, then $Q_u/Q_s = 1.130$, i.e. the unsteady flow is 13% greater than that at the equivalent stage in steady flow conditions. In this case the ratio of kinematic wave speed to fluid velocity, c/u, has been taken as $5/3$, the theoretical value for flow in a wide rectangular channel (Henderson, 1966).

IN-BANK FLOWS

Velocity field

The distribution of primary velocity in an open channel with a particular cross section remains an intractable problem due to the non-isotropic nature of the turbulence. (See Chiu & Chiou (1986), Einstein & Li (1958), Liggett, Chiu & Miao (1965), Melling & Whitelaw (1976), Noat & Rodi (1982), Perkins (1970), Rodi (1980), Tracy (1965)). Various attempts have been made to predict the velocity field but as each theoretical turbulence model becomes more sophisticated, so do its requirements for empirical information. For example, Naot and Rodi's algebraic stress model requires 10 empirically derived coefficients and although the results are useful at the research level, they are not yet of sufficient generality to be useful to the design engineer.

A typical pattern of isovels for flow in a straight trapezoidal channel is shown in Figure 6.5. Also shown are measured boundary shear stress values. The aspect ratio, b/h (= base width to depth ratio) was 1.52 and the flow was supercritical with a Froude number, Fr, of 3.24. For further details see Yuen (1987). The isovels clearly indicate the three-dimensional nature of the flow and the influence of secondary flows. At this particular aspect ratio there are clearly two contra-rotating secondary flow cells near each corner between the bed and the walls, convecting high momentum fluid into each corner region. The cell returns flow along the bed towards the centreline, at which point it meets the return flow from the opposite corner. The secondary flow is thus directed upwards at the centreline normally away from the bed. The isovels are thus more spaced out in this region and the local bed shear stress correspondingly reduced. For further details of how patterns of contra-rotating secondary flow cells affect boundary

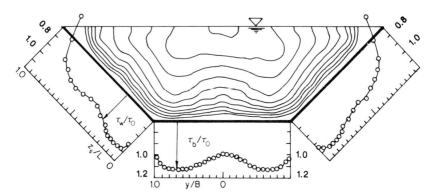

Figure 6.5 Typical relationship between boundary shear stress distribution and primary velocity, Fr = 3.24, Aspect ratio = 1.52 (Yuen, 1987).

shear stress and velocity at various aspect ratios see Knight and Patel (1985a), Ligget et al. (1965), Nikagawa et al. (1983), Naot and Rodi (1982), Nezu and Nakagawa (1984), Odgaard (1984) and Tracy (1965).

An important consequence of the three-dimensional nature of the flow, even in smooth straight prismatic channels, is that the classical logarithmic velocity distribution laws apply only close to the boundary and not over the entire cross-section. For rough channels the influence of secondary flows will be diminished. However, under these circumstances, and particularly for shallow depths over very rough boundaries, the application of the logarithmic law is further complicated by the difficulties in specifying both the bed datum and the appropriate roughness height. It follows from these comments therefore that care should be exercised before using the logarithmic law under all circumstances to determine the local boundary shear stress for sediment transport calculations. Wherever possible checks should be made by measuring either the vertical distribution of Reynolds stress over a vertical (West et al. (1984) or the longitudinal energy gradient (Wallis and Knight, 1984).

One practical consequence for secondary flows in straight channels, especially those with a rectangular cross section, is that the filament of maximum velocity may no longer occur at the centreline of the channel free surface but will probably be depressed below the surface. (Schlichting, 1979, Knight et al. 1984). Changes in plan form geometry, such as bends, also enhance secondary flows and distort the isovel pattern even further. This again highlights the difficulty of predicting the velocity field accurately in natural channels. The spatial distribution of primary velocity should therefore always be measured directly, if accurate values are required.

Boundary shear stress distribution

Because the local boundary shear stress is such an important parameter in sediment mechanics it follows that its distribution around the wetted perimeter of an open channel is helpful in determining local effects. For example, the position and value of the maximum boundary shear stress is of importance in erosion problems, and the mean boundary shear stress on a particular boundary element may be required for accurate predictions of sediment transport rate. It is only in recent years that the measurement of boundary shear stress around the wetted perimeter of laboratory channels and ducts of arbitrary cross-section has been attempted, largely due to a simple device known as the Preston tube (Preston, 1954; Patel, 1965; Winter, 1977). Although this device only gives the temporal mean value of local boundary shear stress, whereas in sediment mechanics the fluctuations about the mean and their correlation with the fluctuating lift forces on individual particles are also required, nevertheless it is a parameter widely used in equations for sediment transport. In natural rivers electromagnetic flowmeters (Brierley *et al.*, 1986) afford an alternative way of measuring bed shear stress.

As already indicated for the velocity field, theoretical turbulence models have severe limitations. Patterns of boundary shear stress distribution therefore normally have to be determined empirically for each shape of channel cross-section. Taking a rectangular cross-section as the simplest shape, Figures 6.6 and 6.7 indicate the complexity of the spanwise distribution of bed shear stress, τ_b, for duct flow (Knight and Patel, 1985b), in which aspect ratio, B/H, is varied

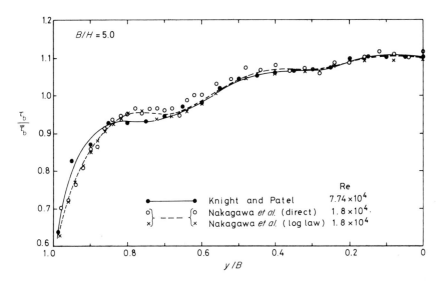

Figure 6.6 Spanwise distribution of bed shear stress, $B/H = 5.0$.

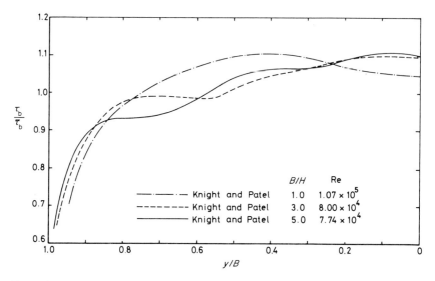

Figure 6.7 Comparison of the spanwise distributions of bed shear stress for different aspect ratios non-dimensionalized by the mean bed shear stress, $\bar{\tau}_b$.

from 1.0, a square cross-section to 5.0 In these 2 figures B and H refer to the semi-width and depth of the duct respectively. Figure 6.6 indicates that despite the criticisms made earlier, the values of τ_b based on the logarithmic velocity law (symbol x agree quite well with those obtained with a Preston tube (symbol •) or a directly calibrated hot film (symbol ○). Figure 6.7 shows how the spanwise distribution is affected by aspect ratio, taking just three particular aspect ratios for the purposes of illustration. It is clear that the local bed shear stress at a given spanwise coordinate can change significantly as a result of quite small changes in aspect ratio. A possible mechanism to explain this is given elsewhere (Knight and Patel, 1985a,b).

In open-channel flow similar trends may be observed. In this case the presence of the free surface makes a small difference in secondary flow cell pattern. By integrating the boundary shear stress distributions vertically and laterally over the wall and bed elements, the mean wall and bed shear stress τ_w and τ_b may be obtained. When multiplied by the appropriate element of wetted perimeter the corresponding shear forces may be obtained. The shear forces acting on the walls and the bed of a rectangular open channel, of total base width b, expressed as percentages of the total shear force per unit length $SF_T (= \tau_o P = \rho g A S_f)$, are given by

$$SF_w = \bar{\tau}_w 2h$$
$$SF_b = \bar{\tau}^b b$$

Hydraulics of Flood Channels

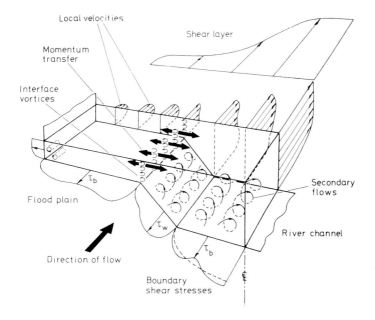

Figure 6.8 Hydraulic aspects of overbank flow.

$$SF_T = SF_w + SF_b$$

$$\%SF_w = 100 SF_w / SF_T. \tag{6.16}$$

Experiments indicate (Knight *et al.*, 1984) that the $\%SF_w$ has an exponential relationship with the aspect ratio b/h, of the form

$$\%SF_w = e^\alpha \tag{6.17}$$

where

$$\alpha = -3.230 \log\left(\frac{b}{h} + 3\right) + 6.146. \tag{6.18}$$

The mean wall and bed shear stresses may then be determined using the ancillary equations

$$\bar{\tau}_w/(\rho g h S_f) = 0.01\%SF_w(b/2h)$$
$$\bar{\tau}_w/(\rho g R S_f) = 0.01\%SF_w(1 + b/2h)$$
$$\bar{\tau}_b/(\rho g h S_f) = 1 - 0.01\%SF_w$$
$$\bar{\tau}_b/(\rho g R S_f) = (1 - 0.01\%SF_w)(1 + 2h/b) \tag{6.19}$$

96   Floods

Similar equations apply to trapezoidal channels (Yuen, 1987) and rectangular closed ducts (Knight *et al.*, 1982). For other shapes further experimental work is required, although for shallow parabolic cross-sections, typical of many rivers, it is often assumed that τ_b varies across the width according to the local depth of flow (Henderson, 1966, page 452).

Sediment transport

The river engineer is generally concerned not only with conveyance capacity, but also with the stability of the channel in both plan form and shape, erosion and deposition, threshold criteria for sediment movement, armouring, sediment transport capacity, resistance to flow and many other related topics. Each of these phenomena depends significantly upon the local boundary shear stress and the comments made earlier will therefore apply. For example, secondary flow cells may produce longitudinal ribs in sand bed channels at low Reynolds numbers (Ikeda, 1981), the distribution of bed shear stress governs the 'active' width of a river channel and the maximum boundary shear stress is likely to be a limiting parameter in channel design. These last two features may be illustrated by reference to Figure 6.7. Taking $\bar{\tau}_b$ as a representative threshold shear stress, then for an active bed $\tau_b/\bar{\tau}_b$ must be greater than unity. The curves for $B/H = 1.0$ and 3.0 indicate that the active width decreases from around 80% to 50% for this small change in aspect ratio. Similarly Figure 6.7 shows that the peak bed shear occurs as the centreline for $B/H = 3.0$ but at $y/B = 0.4$ for $B/H = 1.0$. Figure 6.5 further shows that τ_b/τ_o is a maximum at $y/B = 0.65$ and that τ_w/τ_o is a maximum at $z_s/l = 0.15$.

However, for floods that remain in-bank, such as might occur in a well-trained river or narrow valley, it should be recognised that under extreme conditions a great deal of flood damage can be caused by mud slides, erratic boulders, debris and other material flowing in the water. The mechanism whereby large boulders can slide on a carpet of smaller stones is also an impotant feature of bank and bed protection.

OUT-OF-BANK FLOWS

Hydraulic aspects of overbank flow

The stage–discharge curve in Figure 6.1(a) and the standard procedure for calculating normal depth flow both indicate a relatively simple relationship between depth of flow and discharge. However, in most natural rivers, as the discharge rises the depth of flow or stage will also increase until the water level is coincident with the level of the channel banks. The river is then said to be flowing at bankfull, and the discharge is often referred to as the bankfull

Figure 6.9 Compound channel section of the R. Main, N. Ireland (courtesy of Dr. W. R. C. Myers).

discharge, Q_b. Any further increase in discharge implies that the river will flow in an out-of-bank condition flooding the adjacent land or flood plains. The bankfull discharge is therefore an important parameter both from the land drainage and the geomorphological point of view. Although the return period of the bankfull discharge is frequently adopted in design studies, the limitations behind its use should be appreciated (Lewin, 1989; Nixon, 1959). A discontinuity will inevitably occur in the stage–discharge curve at bankfull level due not only to the effects of channel shape referred to earlier, but also to the hydraulic features shown in Figure 6.8.

This figure shows that once a channel flows in an out-of-bank condition the generally slower moving water on the flood plain retards the faster flowing water in the river channel, thereby creating a shear layer which extends laterally some considerable distance both into the main river channel and onto the flood plain. This shear layer is indicative of some transfer of momentum between the flood plain flow and the main river channel flow and is shown schematically in Figure 6.8 as taking place through a bank of interface vortices rotating on vertical axes. It is this momentum transfer which further complicates the already three-dimensional flow structures occurring in the in-bank channel flow and makes the stage–discharge curve difficult to predict for the out-of-bank condition. Many studies have been undertaken into this type of channel flow, and a comprehensive literature review (Hollinrake, 1987) has recently been completed. Computational techniques concerning overbank flow in flood routing models are provided in specialist texts (e.g. Cunge *et al.*, 1980; Samuels, 1984). An example of a compound or two-stage channel is shown in Figure 6.9, which shows a canalized section of the River Main in Northern Ireland flowing below bank full.

LATERAL MOMENTUM TRANSFER

A typical set of overbank flow data are shown in Figure 6.10. The data are from smooth duct flow experiments (Lai, 1987) with B/b = 4.94, b/h = 1.0 and $(H-h)/H$ = 0.134 where B now refers to the top half width of the channel (see Figure 6.11). The notation is defined in Figure 6.11. The lateral spread of the shear layer on the flood plain is clearly apparent, extending to nearly 50% of the flood plain width. The values of boundary shear stress (x), depth mean velocity (Δ), momentum (+) and kinetic energy (Δ) all level off to their nominal two-dimensional value before decreasing in the corner region. The shear layer in the main channel is not fully developed due to the low aspect ratio of the main channel ($2b/h$ = 2.0). The influence of secondary flows in the main channel are particularly noticeable in the isovel pattern and the bed shear stresses. It is interesting to note that the depth mean velocity remains sensibly constant over the centre half of the channel, although the bed shear stress varies significantly.

The measured boundary shear stress values may be integrated along each boundary element. The difference between the weight force resolved down the channel bed slope and this shear force gives the apparent shear force, ASF, on any internal interface drawn at that lateral position. Thus referring to Figure 6.11, a force balance for the flow in the main channel gives

$$2ASF_v + 2SF_3 + SF_4 = \rho g 2bHS_f$$

$$2ASF_H + 2SF_3 + SF_4 = \rho g 2bhS_f$$

$$2ASF_I + 2SF_3 + SF_4 = \rho g b(h + H)S_f \qquad (6.20)$$

where ASF_V, ASF_H and ASF_I are the apparent shear forces on vertical, horizontal and inclined interfaces respectively. Provided SF_3 and SF_4 are measured or known then the apparent shear forces, ASF, or apparent shear stresses, ASS, may be determined. They are usually expressed as a percentage of the total channel shear force (= $\tau_o P$), to give $\%ASF_V$, $\%ASF_H$ and $\%ASF_I$ values. Experimental results (Knight and Demetriou 1983; Knight and Hamed, 1984) indicate that the $\%ASF_V$ value for a single vertical interface can exceed 25% of the total channel shear force for rough flood plains, and may be in excess of 10% even for smooth channels. This further illustrates the significance of the momentum transfer between different sub-areas of a river flowing overbank. It also illustrates the general problem of dealing with open-channel flows where significant lateral changes in either depth or roughness occur, as highlighted earlier.

Conveyance capacity

The discharge capacity of a compound channel is seriously reduced at low relative depths ($0 < (H-h)/H < 0.2$) and large flood plain widths due to the momentum transfer between the regions of slow and fast moving fluid. In certain cases the conveyance capacity of a compound channel may in fact be actually less than that of the main channel at the same stage but in a non-interacting mode (Knight and Hamed, 1984). This means that particular care should be taken in estimating either level or discharge when channels are just flowing out-of-bank, and especially at relative depths of around 0.1, which corresponds to the stage at which the apparent shear force and therefore the momentum exchange is at a maximum. It also implies that in unsteady flow the stage–discharge relationship will experience some hysteresis in the vicinity of bankfull stage and care should be exercised in calibrating flood-routing models.

Design procedures for the conveyance capacity of compound channels usually rely on splitting the flow area into a number of sub-areas before applying standard resistance laws, such as those given by equations (6.3)–(6.5). Such

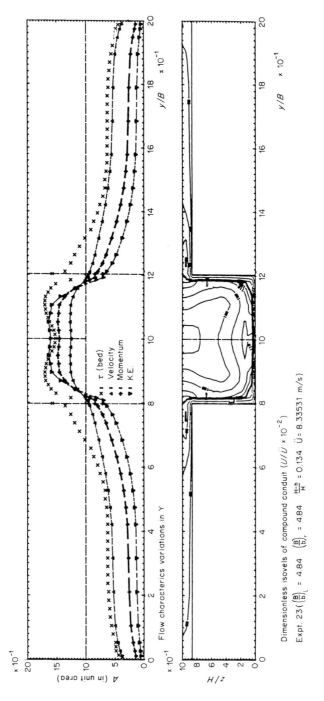

Figure 6.10 Typical set of overbank data for duct flow (Lai, 1987).

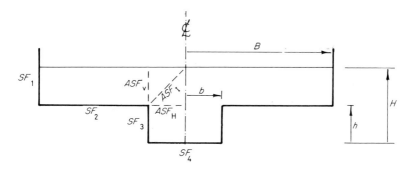

Figure 6.11 Notation for sub division of a compound channel.

procedures will give erroneous answers unless the apparent shear forces on internal interfaces are included. Particular care should also be attached to obtaining the correct proportion of flow between sub-areas (Myers, 1987; Wormleaton and Hadjipanos, 1985). The proportion of discharge in the main channel is usually significantly larger than that based on a simple proportioning of areas between the main channel and the total flow area, i.e. the main channel and and flood plain areas combined. At large relative depths ((H-h)/H > 0.5) the flow and geometric proportions are much closer (Knight and Demetriou, 1983).

Further details of flow structure are of little importance in one-dimensional models, apart from the kinetic energy and momentum correction coefficients, α and β, and the concept of critical depth, h_c. The latter may be of some importance in determining hydraulic controls in steep valleys or in determining longitudinal hydraulic profiles in a dam-break type analysis. Because of the non-uniformity in the velocity field in out-of-bank flow, the α and β coefficients are much larger than those for in-bank flow, and may reach values of 1.5–2.0 and 1.2–1.5 respectively.

The SERC flood channel facility

Because of the difficulty in undertaking detailed measurements in natural flood channels under unsteady flow conditions considerable reliance is, and will continue to be, placed on laboratory data for the purposes of calibrating mathematical models. Much of this data has been obtained in university laboratory channels. In November 1985 the Science and Engineering Research Council (SERC) commissioned a flood channel facility in order to coordinate work in this area on one large central facility. The facility consists of a 56 m long by 10 m wide flood channel, with a discharge capacity of 1.1 m^3/s (see Figure 6.12).

Figure 6.12 The SERC Flood Channel Facility (courtesy of Hydraulics Research Ltd, Wallingford).

In addition to traditional instrumentation, such as miniature current meters, water level recorders and Preston tubes, the facility is also equipped with a sophisticated two-component laser anemometry system. The laser anemometer has a miniaturised cylindrical optical head, 15 mm in diameter and 100 mm long, which can be placed at any location with the cross-section of the facility. The probe head is coupled by 20 m of armoured fibre-optic cable to the laser and transmission optics. The probe senses two velocity components simultaneously in the plane normal to the axis of the probe head and thus produces sufficient information to calculate the turbulent Reynolds stresses.

A five-year programme of research, extending from 1986 to 1990, has been formulated by seven UK university research teams and practising engineers who together comprise the SERC Working Party on Flood Channels. The research programme includes studies of straight, skewed and meandering channels each flowing in an out-of-bank condition. Additional studies on dispersion, geomorphology and sediment transport are also planned. A fuller description of the facility and the first phase of the research programme is given elsewhere (Knight and Sellin, 1987).

CONCLUDING REMARKS

The hydraulic behaviour of flood channels is clearly not a simple matter. The fluid flow is essentially three-dimensional, unsteady and may interact with the boundary. The geomorphological implications of this are that care should be exercised when attempting to quantify the link between either surface erosion or deposition and the fluid flow. However, as our knowledge of the flow mechanisms, velocity field, boundary shear stress, mass and momentum transfer, eddy viscosity and Reynolds stress distributions increases, so the predictive capability of flow models or the interpretation of geomorphological phenomena will improve.

The particular problem of out-of-bank flow is a common feature in fluvial systems and of great practical significance in the design of flood alleviation works. The lateral transfer of momentum between the main river channel and the flood plains has been identified as an important phenomenon because it affects not only the distribution of flow parameters but also the sediment fluxes of suspended load onto and off the flood plains. The overall conveyance capacity of a compound channel has been shown to be difficult to determine accurately due to the discontinuity in shape, hydraulic radius and possibly roughness at the main channel/flood plain interface. These combine to significantly affect the stage discharge curve, momentum transfer and friction factors at low flood plain depths.

REFERENCES

ASCE (1963). Friction factors in open channels, Task Force Report, *J. Hyd. Div.*, ASCE, **89**, HY2, 97–143.

BHRA (1983). *Hydraulic Aspects of Floods and Flood Control*, 1st Int. Conference, British Hydromechanics Research Association, Bedford, UK.

BHRA (1985). *The Hydraulics of Floods and Flood Control*, 2nd Int. Conference, British Hydromechanics Research Association, Bedford, UK.

Brierley, R. W., Shiono, K., and West, J. R. (1986). An integrated system for measuring estuarine turbulence, *Proc. Int. Conf. Measuring Techniques of Hydraulics Phenomena in Offshore, Coastal and Inland Waters*, BHRA, Cranfield, 359–76.

Chiu, C. L., and Chiou, J. D. (1986). Structure of 3-d flow in rectangular open channels, *J. Hyd. Engrg*, ASCE **112** (HY11), 1050–68.

Chow, V. T. (1959). *Open Channel Hydraulics*, McGraw-Hill, New York.

Cunge, J. A., Holly, F. M., and Verwey, A. (1980). *Practical Aspects of Computational River Hydraulics*, Pitman, London.

Einstein, H. A., and Li, H. (1958). Secondary currents in straight channels, *Trans. American Geophysical Union*, **39**(6), 1085–8.

Ervine, D. A., and Ellis, J. (1987). Experimental and computational aspects of overbank floodplain flow, *Trans. Roy. Soc. Edinburgh: Earth Sciences*, **78**, 315–25.

French, R. H. (1986). *Open Channel Hydraulics*, McGraw-Hill, Singapore.

Garde, R. J., and Ranga Raju, K. G. (1977). *Mechanics of Sediment Transportation in Alluvial Streams*, Wiley, New Delhi.

Henderson, F. M. (1966). *Open Channel Flow*, Macmillan, New York.
Herschy, R. W. (1978). *Hydrometry*, Wiley, Chichester.
Hey, R. D., Bathurst, J. C., and Thorne, C. R. (1982). *Gravel Bed Rivers*, Wiley, Chichester.
Hollinrake, P. G. (1987). *The Structure of Flow in Open Channels-a Literature Search*, Hydraulics Research Ltd. Report, SR96, Wallingford, UK, 327 pp.
Ikeda, S. (1981). Self formed straight channels in sandy beds, *Journal of Hydraulics Division*, ASCE, **107**(HY4), 389-406.
ISO (1982). Liquid flow measurement in open channels—Part 2: Determination of the stage discharge relation, *Int. Organisation for Standardisation*, ISO 1100/2, 1-33.
Jones, O. C. (1976). An improvement in the calculation of turbulent flow in rectangular ducts, *J. Fluids Engng., Trans. ASME*, June, 173-81.
Knight, D. W. (1981). Boundary shear in smooth and rough channels, *J. Hyd. Div., ASCE*, **107**(HY7), 839-51.
Knight, D. W. (1985). Advances in river engineering in T. H. Y. Tebutt, (ed.), *Advances in Water Engineering*, Elsevier, London, pp. 278-88.
Knight, D. W., and Demetriou, J. D. (1983). Flood plain and main channel flow interaction, *Journal of Hydraulic Engineering*, ASCE, **109**(HY8), 1073-2.
Knight, D. W., Demetriou, J. D., and Hamed, M. E. (1984). Boundary shear in smooth rectangular channels, *Journal of Hydraulic Engineering*, ASCE, **110**(4), 405-22.
Knight, D. W., and Hamed, M. E. (1984). Boundary shear in symmetrical compound channels, *Journal of Hydraulic Engineering*, ASCE, **110**(10), 1412-30.
Knight, D. W., and Patel, H. S. (1985a). Boundary shear in smooth rectangular ducts, *Journal of Hydraulic Engineering*, ASCE, **111**(1), 29-47.
Knight, D. W., and Patel, H. S. (1985b). Boundary shear stress measurements in rectangular duct flow, *2nd Int. Symp. on Refined Flow Modelling and Turbulence Measurements*, Iowa, USA.
Knight, D. W., Patel, H. S., Demetriou, J. D., and Hamed, M. E. (1982). Boundary shear stress distributions in open channels and closed conduit flows, *Proceedings of Euromech 156-The Mechanics of Sediment Transport*, Istanbul, July, A. A. Balkema, Rotterdam, Netherlands, 33-40.
Knight, D. W., and Sellin, R. H. J. (1987). The SERC flood channel facility, *Journal of the Institution of Water and Environmental Management*, IWEM, **1**(2), 198-204.
Kouwen, N., and Li, R. M. (1980). Biomechanics of vegetative channel linings, *J. Hyd. Div., ASCE*, **106**(HY6), June, 1085-1103.
Lai, C. J. (1987). Flow resistance, discharge capacity and momentum transfer in smooth compound closed ducts, *PhD Thesis*, University of Birmingham, UK.
Larsen, P. (1973). Hydraulic roughness of ice covers, *Journal of Hydraulics Division*, ASCE, Vol. 99, No. HY1, January, 111-119.
Lewin, J. (1989). Floods and fluvial geomorphology, *The Hydrology, Sedimentology and Geomorphological Implications of Floods*, K. J. Beven and P. Carling, (eds), J. Wiley, London.
Liggett, J. A., Chiu, C. L., and Miao, L. S. (1965). Secondary currents in a corner, *Journal of the Hydraulics Division*, ASCE, Vol. 91, No. HY6, November, 99-117.
Melling, A., and Whitelaw, J. H. (1976). Turbulent flow in a rectangular duct, *Journal of Fluid Mechanics*, Vol. 78, part 2, 289-315.
Myers, W. R. C. (1987). 'Velocity and discharge in compound channels', *Journal of Hydraulic Engineering*, ASCE, Vol. 113, No. 6, June, 753-66.
Nakagawa, H., Nezu, I., and Tominaga, A. (1983). Secondary currents in a straight channel flow and the relation to its aspect ratio, *4th Int. Symp. on Turbulent Shear Flows*, Karlsruhe.

Naot, D. (1984). Response of channel flow to roughness heterogeneity, *J. Hyd. Engng.*, ASCE, **110**(11), 1568–587.

Naot, D., and Rodi, W. (1982). Calculation of secondary currents in channel flow, *Journal of the Hydraulics Division*, ASCE, **108**(HY8), August, 948–68.

Nezu, I., and Nakagawa, H. (1984). Cellular secondary currents in straight conduit, *Journal of Hydraulic Engineering*, ASCE, **110**(2), February, 173–93.

Nixon, M. (1959). A study of bank-full discharges of rivers in England and Wales. *Proc. Inst. Civil Engineers*, London, Paper 6322, February, 157–74.

Odgaard, A. J. (1984). Shear-induced secondary currents in channel flows. *Journal of Hydraulic Engineering*, ASCE, **110**(7), July, 996–1004.

Patel, V. C. (1965). Calibration of Preston tube and limitations on its use in pressure gradients, *Journal of Fluid Mechanics*, **23**, Part I, 185–208.

Perkins, H. J. (1970). The formation of streamwise vorticity in turbulent flow, *Journal of Fluid Mechanics*, **44**, Part 4, 721–740.

Preston, J. H. (1954). The determination of turbulent skin friction by means of Pitot tubes, *Journal Roy. Aero. Soc.*, **58**, 109–121.

Price, R. K. (1985). Hydraulics of floods, in *Advances in Water Engineering*, T. H. Y. Tebutt, (ed.), Elsevier, London, 302–10.

Ree, W. O., and Palmer, V. J. (1949). Flow of water in channels protected by vegetative linings, *U.S. Soil Conservation Service, Technical Bulletin 967*, February, USA.

Rehme, K. (1973). Simple method of predicting friction factors of turbulent flow in non-circular channels, *Int. J. Heat and Mass Transfer*, **16**, 933–50.

Rodi, W. (1980). *Turbulence Models and their Application in Hydraulics*, State of the Art paper, IAHR, Netherlands, 104 pp.

Samuels, P. G. (1984). *Computational Modelling of Open Channel Flow–An Analysis of Some Practical Difficulties*, Hydraulics Research Ltd. Report No. IT 273, Wallingford, UK.

Schlichting, H. (1979). *Boundary Layer Theory*, McGraw-Hill, New York.

Shen, H. W. (1979). *River Mechanics*, Colorado State University, USA.

Smith, K. V. H. (1984). *Channels and Channel Control Structures*, 1st Int. Conf., Springer-Verlag, Berlin.

Subramanya, K. (1982). *Flow in Open Channels*, Vols. I & II, Tata McGraw-Hill, New Delhi.

Tracy, H. J. (1965). Turbulent flow in a three-dimensional channel, *Journal of the Hydraulics Division*, ASCE, **91**(HY6), 9–35.

Vanoni, V. A. (1975). *Sedimentation Engineering*, M. & R No. 54, ASCE, New York.

Wallis, S. G., and Knight, D. W. (1984). Calibration studies concerning a one dimensional numerical tidal model with particular reference to resistance coefficients, *Estuarine, Coastal and Shelf Science*, Academic Press, **19**, 541–62.

West, J. R., Knight, D. W., and Shiono, K. (1984). A note on flow structure in the Great Ouse estuary, *Estuarine, Coastal and Shelf Science*, Academic Press, **19**, 271–90.

Winter, K. G. (1977). An outline of the techniques available for the measurement of skin friction in turbulent boundary layers, *Prog. Aerospace Sci.*, Pergamon, **18**, 1–57.

Wormleaton, P. R., and Hadjipanos, P. (1985). Flow distribution in compound channels, *Journal of Hydraulic Engineering*, ASCE, **111**(2), 357–61.

Yuen, K. W. H. (1987). *A Study of Boundary Shear Stress and Flow Resistance in Open Channels with a Trapezoidal Cross Section*, M.Sc. Qualifying Thesis, University of Birmingham, UK.

7 Flow-Competence Evaluations of the Hydraulic Parameters of Floods: an Assessment of the Technique

PAUL D. KOMAR
College of Oceanography, Oregon State University, Corvallis, Oregon

INTRODUCTION

The concept of flow competence was introduced by Gilbert and Murphy (1914) to refer to the limiting case of no sediment transport by flowing water. In their application flow competence did not differ from our standard condition for the threshold of uniformly sized grains by a current since such deposits were used in their flume experiments. However, Gilbert and Murphy did make the comment (p. 35): 'A current flowing over debris of various sizes transports the finer but can not move the coarser; the fineness of the debris it can barely move is the measure of its competence.' Although there was initial confusion as to its meaning, flow competence has evolved into a measure of a current's sediment-moving ability inferred from the largest particles transported under a given set of flow conditions (Baker and Ritter, 1975; Costa, 1983). The assessment of competence has thereby become a useful tecnique for the evaluation of extreme floods in still-active river systems. For example, Bradley and Mears (1980) analyzed the flood velocities and bed stresses responsible for the transport of boulders up to 2.3 meters diameter found during excavations adjacent to Boulder Creek where it flows through Boulder, Colorado. Their application is typical of those for still-active rivers, the analyses examining extreme flood events that are often prehistoric and beyond the normal range of experience; one is conceivably evaluating the hydraulic parameters of the 500-year or 1000-year flood. Examples of competence evaluations of still larger and more ancient floods are provided by Baker (1973) and Lord and Kehew (1987), extreme floods produced by glacial-lake outbursts. The technique can also be applied to river-flood deposits found in the geologic rock record. Some of the formulae presented in this paper are not restricted in use to coarse-sediment movement by rivers, and could be employed in any environment of transport by flowing water, modern or ancient.

The quotation from Gilbert and Murphy (1914) recognizes that flow competence is an aspect of selective grain entrainment from a bed of mixed sizes.

That it does involve such processes is obvious, but this aspect has been largely ignored in the development of equations for competence evaluations. However, recently obtained data for selective entrainment of gravel have forced this recognition. Those measurements indicate that our past evaluations of competence may have been seriously in error, typically having over-estimated bed stresses, velocities and discharges of floods.

I have undertaken comparisons between the competence formulae and empirical relationships based on selective-entrainment measurements (Komar, 1987a). Other analyses focused on the physical processes of flow drag and lift which entrain a clast from a bed of generally smaller grains, leading to theoretical relationships that can be applied to competence evaluations (Komar, 1988). A brief review of those analyses will be presented here together with speculations on the practical aspects of their applications. Comparisons will also be made with formulae developed by J. Bathurst and co-workers (Bathurst, *et al.*, 1983, 1987; Bathurst 1987) that yield the flow discharge required for entrainment, formulae that could also be employed in competence evaluations. Another objective of this overview of competence evaluations is to examine factors that might affect the assessments, factors such as the distribution of flow stresses on the stream bed, the effects of steep channel gradients, and the entrainment of large boulders which project above the water level or comprise a significant portion of the channel cross-section. Examples of the application of flow-competence evaluations will be presented in the final section of this paper.

FLOW COMPETENCE AND SELECTIVE ENTRAINMENT

Novak (1973), Baker and Ritter (1975) and Costa (1983) compiled data for the transport of extreme clast sizes in attempts to establish empirical relationships for the evaluation of flow competence. The data do yield a strong trend of increasing mean-flow or velocity with increasing clast size. Based on data for diameters spanning the range 1–500 cm, Costa obtained a flow-competence relationship equivalent to

$$\tau_c = 26.6 D^{1.21} \tag{7.1}$$

where τ_c is the mean-flow stress having units dynes/cm^2 and D is the clast diameter in cm (the units have been changed from those used by Costa). It is such a formula that traditionally has been used for competence evaluations, providing a seemingly simple and direct evaluation of the flood's mean stress from the largest-diameter clast transported (the average of the five largest clasts is often used rather than the absolute maximum size, thereby providing a more representative transport competence).

The data used to establish such competence formulae come from a variety of sources. Some measurements are derived from actual floods, yielding information on the maximum sizes of material transported. However, during floods it is not generally possible to directly measure the entrainment flow stress, τ_c, so this is estimated from the DuBoys relationship $\tau = \rho g h S$, where ρ is the density of water, g is the acceleration of gravity, h is the flow depth or hydraulic radius, and S is the channel slope or energy gradient. Therefore, flow stresses were calculated from measured or inferred flow depths and channel gradients, employing a relationship which assumes uniform-flow conditions. Other data used to establish the competence relationship come from more controlled conditions, an example being the study of Helley (1969) where cobbles were placed on the bed and their movement monitored under the naturally varying stream discharge. The measurements of Milhous (1973) have also been used, a study which employed a bedload trap to capture the moving gravel.

I have undertaken a detailed review of the data used to establish equation (7.1) (Komar, 1987a). My primary conclusion was that although the combined measurements do yield a strong trend of increasing τ_c with increasing D and accordingly yield equation (7.1), the individual data sets either follow no statistically significant trends or have trends which actually run counter to the overall flow-competence relationship. In addition, individual data sets which do have statistical trends yield relationships which can be accounted for by considerations of selective entrainment from deposits of mixed grain sizes. This in part led me to suggest that competence relationships such as equation (7.1) be abandoned in favor of direct examinations of selective-entrainment processes and formulae derived from those analyses.

Data on selective entrainment are illustrated by the measurements of Carling (1983, 1987a) who employed bed-load traps in gravel-bed streams of the Pennines in England. Carling also tagged still larger clasts (up to 44 cm diameter), and monitored the flow conditions required for their entrainment. A correlation between mean flow stresses and the maximum-size clasts transported in Great Eggleshope Beck yielded the relationship

$$\tau_{ti} = 110 D_{bi}^{0.38} \qquad (7.2)$$

where the stress and diameter units are again dynes/cm² and cm. The i subscript is used to denote that this is a threshold relationship for the selective entrainment of individual grains from a deposit of mixed sizes. The b subscript indicates that in this case the grain size is the intermediate axial diameter.

A similar correlation is obtained in analyses of the data obtained by Hammond *et al.* (1984) for gravel entrainment under tidal currents on the continental shelf off the south coast of England. Grain movements was observed with an underwater TV video and the velocity profile was measured with electromagnetic current meters, the gradient of the velocity profile in turn yielding the mean flow stress. Although the environment is much different than the rivers investigated by Carling (1983), the resulting relationship

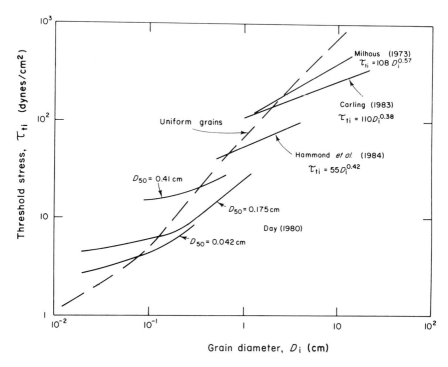

Figure 7.1 A compilation of selective-entrainment curves for grain movement from deposits of mixed sizes, compared with the standard threshold curve (dashed) for the threshold of uniform grains (From Komar, 1987b).

$$\tau_{ti} = 55 D_{bi}^{0.42} \tag{7.3}$$

is similar. The primary difference is the proportionality coefficients (110 versus 55), but this coefficient reflects the overall grain-size range involved in the studies, having been coarser in the river investigated by Carling. More interesting are the exponents (0.38 versus 0.42) as these reflect the degree of selective entrainment, the lower the value the greater the degree of sorting by size under varying flow conditions (the greater the change in D_{bi} for a given variation in τ). The near equality of the exponents in equations (7.2) and (7.3) indicates comparable degrees of selective entrainment and sorting in the two data sets even though they are from considerably different environments.

A compilation of such selective-entrainment relationships is given in Figure 7.1, including curves based on the data of Milhous (1973) obtained in a gravel-bed stream and the laboratory-flume measurements of Day (1980). Details of the analyses of these data can be found in Komar (1987b). The dashed line is that based on laboratory-flume measurements where the experiments were

Flow-Competence Evaluations of the Hydraulic Parameters of Floods

performed with essentially uniform gains. Those data have been summarized by Miller et al. (1977) and Yalin and Karahan (1979); both show that with beds of coarse grains and high-grain Reynolds numbers, the Shields entrainment function is approximately constant at $\theta_t = \tau_t/(\rho_s-\rho)gD \approx 0.045$ (rather than the 0.06 value originally given by Shields); ρ_s and ρ are respectively the grain and water densities. This yields

$$\tau_t = 0.045(\rho_s - \rho)gD \tag{7.4}$$

for the threshold stress for deposits of uniformly sized sediments, the straight portion of the dashed curve in Figure 7.1 for grain diameters larger than about 0.1 cm.

Each of the selective-entrainment curves in Figure 7.1 is seen to obliquely cross the dashed curve for uniform grains, the exponents of those relationships being less than the $\tau_t \propto D$ proportionality of equation (7.4). This is reasonable in that within a deposit of mixed sizes, the larger grains will be able to move more readily than if part of a deposit of that size alone, while due to sheltering effects, etc., the smaller grains will be more difficult to entrain than when they form a uniform deposit. The cross-over point of a selective entrainment curve and the curve for uniform grains occurs approximately at the median diameter of the deposit of mixed sizes (Komar, 1987b). When the selective-entrainment data are plotted on a Shields diagram, the curves again obliquely cross the standard curve for uniform grains, sloping downward to the right, the lowest θ_{ti} corresponding to the grain Reynolds number of the largest D_i (see Komar (1987b, Fig. 4) and Figure 7.9 later in this paper). Thus the patterns of selective entrainment from deposits of mixed sizes appear to be reasonably well established, at least for coarse-grained sediments.

Figure 7.2 compares the selective-entrainment curves with the flow-competence line of equation (7.1). The competence line roughly parallels that for uniform grains, but is displaced downward to lower flow stresses for a given grain diameter. Like the curve for uniform grains, the competence line traverses the series of selective-entrainment curves. Here I have added the empirical curve based on the data of Fahnestock (1963) which, like the data of Milhous (1973), were used as part of the compilation in establishing the competence relationship, equation (7.1). It can be seen that the trend of the competence relationship results from the echelon arrangement of the series of selective-entrainment curves. The relationship of the curves suggests that the competence equation (7.1) has in part included the effects of selective entrainment, being displaced towards the coarser ends of those curves away from the curve for uniform grains. However, this partial accounting of selective entrainment is inadequate, as will be demonstrated by examples of applications later in this paper.

At this juncture it might be noted that semantic problems have developed in distinguishing between 'flow competence' and 'selective entrainment'. This is

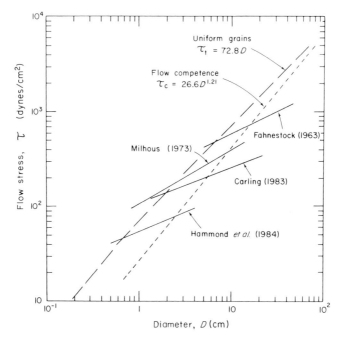

Figure 7.2 A comparison between the flow-competence relationship of equation (7.1) [the short-dashed line], the line from equation (7.4) for the threshold of uniform grains [long-dashed line], and a series of selective entrainment curves (From Komar, 1987a).

illustrated by the data of Carling (1983) which relates the bed stress exerted by the river to the largest grains the flow can entrain and transport. This is a process of selective entrainment, yet at the same time precisely fits the definition of flow competence. Hence, the resulting equation (7.2) derived from Carling's data is a flow-competence relationship as well as one for selective entrainment. However, equation (7.2) is applicable only to Great Eggleshope Beck in the Pennines, while equation (7.3) is still another flow-competence relationship, but one that is suitable only for the gravel deposits studied by Hammond *et al.* (1984). It is obviously desirable to replace these site-specific equations with a more general relationship that is applicable to selective-entrainment and flow-competence evaluations.

EMPIRICAL EVALUATIONS OF SELECTIVE-ENTRAINMENT AND FLOW COMPETENCE

A reasonably generalized relationship is obtained by normalizing the data sets to their respective D_{50} median diameters. Such a relationship for the Shields entrainment function is

Flow-Competence Evaluations of the Hydraulic Parameters of Floods

$$\theta_{ti} = a(D_{bi}/D_{50})^b \tag{7.5}$$

which is dimensionally homogeneous so that the empirical a and b coefficients are dimensionless. The data of Milhous (1973), Carling (1983) and Hammond et al. (1984) are combined in Figure 7.3, and it is seen that they converge reasonably well when normalized by their median grain diameters; the plot further suggests the use of $a = 0.045$ and $b = -0.65$ as coefficients in equation (7.5). This value for a is what one expects if the cross-over points of the selective-entrainment curves and the Shields curve for uniform grains correspond to the $D_{bi}/D_{50} = 1$ condition. The b exponent again reflects the degree of selective entrainment ($b = -1$, a 45° slope, is the limiting value, but otherwise, the greater the slope the greater the degree of sorting).

By definition $\theta_{ti} = \tau_{ti}/(\rho_s - \rho)gD_{bi}$, so that equation (7.5) can be modified to

$$\tau_{ti} = 0.045(\rho_s - \rho)gD_{50}^{0.65}D_{bi}^{0.35} \tag{7.6}$$

if one uses $a = 0.045$ and $b = -0.65$; the combined data sets are given in this form in Figure 7.4. This gives us a general equation for selective-entrainment and competence evaluations, one that is dimensionally homogeneous and accounts

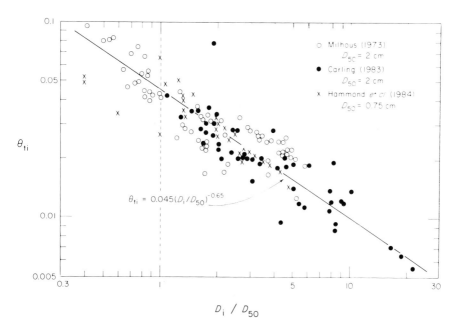

Figure 7.3 The selective-entrainment data of Milhous (1973), Carling (1983) and Hammond et al. (1984) compared with equation (7.5), yielding coefficients $a = 0.045$ and $b = -0.65$ in that relationship.

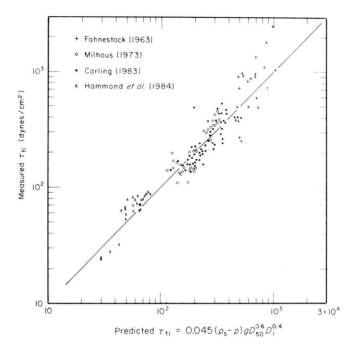

Figure 7.4 Comparisons between measured τ_{ti} mean bed stresses and values predicted with equation (7.6), the line being that for perfect agreement (From Komar, 1987a).

for particle density as well as the median diameter of the deposit as a whole, factors that were not included in equation (7.1). Equation (7.6) reduces approximately to the earlier τ_{ti} versus D_{bi} site-specific relationships; for example, taking $D_{50} = 2$ cm and $\rho_s = 2.7$ g/cm^3 for the gravel in Great Eggleshope Beck studied by Carling (1983), it reduces to $\tau_{ti} = 118 D_{bi}^{0.35}$, which is effectively the same as equation (7.2).

At present there are limits within which equations (7.5) and (7.6) should be applied. They are consistent with the data, up to at least $D_{bi}/D_{50} = 22$ achieved in the measurements of Carling (1983), but might be expected to fail at very large D_{bi}/D_{50}. Although large individual sizes are represented in these data, in all cases the transported clasts were still a part of the basic distribution of bed material. The relationships probably will not apply to extreme D_{bi}/D_{50} ratios where the D_{bi} is an 'erratic', a term employed by Krumbein and Lieblein (1956) to signify extreme grain sizes that are not part of the prevalent size distribution. The entrainment of erratics is complex in that there is usually active transport of the much smaller bed material before the flow is capable of moving the erratic. Entrainment then depends on whether scour around the erratic causes its burial before the flow can roll it along the channel. Fahnestock and Haushild (1962) experimented with clasts having diameters ranging from 3 to 15 cm moving over sand beds. They found that the clasts would move downstream only if the flow was in the upper regime, that is, where the bed forms were plane bed or antidunes. With the lower-flow regime, the clasts were consistently buried within scour holes that developed around them.

Some mention is required of data sets that were not included in the above analyses leading to equations (7.5) and (7.6). Parker et al. (1982), Andrews (1983) and Andrews and Erman (1986) have also compared gravel transport measurements in rivers to equation (7.5), but obtained markedly different coefficients than given above. Of interest is that Parket et al. found $a = 0.0876$ and $b = -0.982$ from a re-analysis of the data of Milhous (1973). Our constrasting results in analysing the data of Milhous are understandable in that Parker et al. were examining the relative transport rates of the different grain-size fractions in the gravel-bed stream, and they utilized equation (7.5) as a normalization parameter for their transport relationships. Therefore, the application by Parker et al. is entirely different than ours which uses the compilation by Milhous of maximum-size clasts transported under the different flow discharges. The results of Parker et al. and those undertaken in the present analyses are not really comparable.

Andrews (1983) derived his data from rivers in Wyoming and Idaho, obtaining $a = 0.834$ and $b = -0.872$ in equation (7.5). Unfortunately, the largest particles captured in the bed-load traps were not measured, so Andrews instead had to employ the geometric mean diameter of the largest sieve-size fraction of material collected. Andrews and Erman (1986) used Helley-Smith bed-load samplers to measure gravel transport in Sagehen Creek, California. My own match of equation (7.5) to their data (Andrews and Erman, Figure 7.7) yields $a = 0.11$ and $b = -1.22$, a surprising result in that this b exponent is equivalent to $\tau_{ti} \propto D_{bi}^{-0.22}$ indicating that there was a decrease in the maximum clast-size transported with increasing flow stress. It appears that this anomalous result can be attributed to their use of Helley-Smith samplers, which had a poor sampling efficiency for the largest bed-load particles whose sizes had intermediate diameters larger than one half the sampler's orifice. This is further indicated by the fact that the largest particles captured ranged up to only 8.6 cm, whereas later experiments with tagged clasts showed that material of 10 cm diameter readily moves even though the floods were weaker than when the Helley-Smith was used. The lesson here is that bed-load samplers may miss the largest clasts in transport so that the results may not be suitable for establishing competence relationships such as equations (7.5) and (7.6). Instead, the determination of maximum-sized particles transported is preferably based on the use of total bed-load traps such as those employed by Milhous (1973) and Carling (1983), or experiments monitoring the movement of large tagged grains.

It has generally been believed that the closer the b exponent of equation (7.5) is to -1, the better the agreement with the equal-mobility hypothesis of Parker et al. (1982). Accordingly, Andrews and Erman (1986) concluded that their measurements confirmed the hypothesis. My own interpretation is that a b = −1 exponent is more apt to represent random data, and has no bearing on the equal mobility hypothesis.

FLOW–DISCHARGE RELATIONSHIPS

Similar analyses have been undertaken by Bathurst *et al.* (1983, 1987) and Bathurst (1987), but following the Schoklitsch approach, which is based on the water discharge per unit flow width rather than on the flow's mean stress or Shields entrainment function. Using flume data for bed materials with relatively uniform sizes, Bathurst *et al.* (1987) obtained the empirical relationship

$$q_t = 0.15 g^{0.5} D^{1.5} S^{-1.12} \tag{7.7}$$

where q_t is the critical water discharge per unit channel width required to entrain grains of diameter D in a channel having a slope S (this equation was developed for particle sizes $0.3 < D < 4.4$ cm and for slopes $0.25 < S < 20\%$). Bathurst (1987) has expanded this approach to an examination of selective entrainment from deposits of mixed sizes. Data from two gravel-bed streams were used to examine the correlation

$$q_{ti} = m D_{bi}^{n}, \tag{7.8}$$

where m and n are empirical coefficients. In plots of q_{ti} versus D_{bi}, the selective-entrainment data were found to have trends which obliquely cross the curve of equation (7.7) for uniform grains, similar to that seen in Figure 7.1 in the τ_{ti} versus D_{bi} comparisons. The cross-over points were again approximately at D_{50} of the distributions, suggesting the normalization of equation (7.8) to

$$q_{ti} = q_t (D_{bi}/D_{50})^k \tag{7.9}$$

where q_t is the value from equation (7.7) for the uniform-grain condition. This relationship is compared with the data of Milhous (1973) in Figure 7.5, where it is seen that there is a good trend yielding

$$q_{ti} = 950 (D_{bi}/D_{50})^{1.22}. \tag{7.10}$$

The scatter in the data decreases at discharges above about 3000 cm^3/(cm.s), which is approximately at the 40 cfs total discharge which Milhous reported as 'critical' for the breakdown of armor stability. Bathurst suggests that the k exponent in equation (7.9) is a function of the D_{84}/D_{16} sorting coefficient of the bed material, obtaining $k = 1.5(D_{84}/D_{16})^{-1}$ based on his river data as well as that for uniform deposits. This predicts that $k \approx 0.5$ for the data of Milhous, but the 1.22 empirical value of equation (7.10) is much closer to the predicted 1.5 value for uniformly sized bed grains. The 950 proportionality coefficient is lower than expected when compared with the predicted q_t from equation (7.7). This 950 value of equation (7.10) is based on $D_{50} = 2$ cm, which is the median for the bed layer as a whole; equation (7.7) predicts that the proportionality coefficient should be 1600. If the analysis is based instead on $D_{50} = 6$ cm for the surface of the armor layer, then the proportionality coefficient of equation (7.10) is increased to 3600 while the predicted value is 8200. From this it would appear that the gravel bed in the stream studied by Milhous was more mobile than in the rivers studied by Bathurst, and that additional investigations are required to establish the empirical coefficients in equation (7.9).

Such problems aside, it is apparent that these unit-width discharge equations from Bathurst and co-workers would potentially be useful in competence evaluations. In some applications it is more desirable to determine flow discharges than bed stresses, so these relationships from Bathurst may be preferred. However, they cannot be viewed as being as general as the τ_{ti} and θ_{ti} equations presented earlier, which can be applied to any water-flow environment: ocean currents and turbidity currents, as well as to river floods. It is also probable that the τ_{ti} and θ_{ti} equations are better suited to large rivers with significant ratios of flow depth to grain sizes, whereas the q_{ti} evaluations may be preferable for smaller streams where the clasts extend above the water level or form a significant part of the mean flow depth since under those conditions it is easier to define flow discharges than mean stresses.

There is no simple relationship between the two approaches, but it is important to establish whether they are generally compatible. The rough-bed velocity-profile equation is

$$\bar{u}(z) = 2.3 \frac{u_*}{\kappa} \log \frac{30z}{k_s} \qquad (7.11)$$

where $u_* = \sqrt{(\tau/\rho)}$ is the shear velocity, z is the vertical distance above the bed, $\kappa = 0.4$ is the von Karman constant, and k_s is the bottom-roughness coefficient. In rivers where this equation can be employed, the depth-integrated discharge is given by

$$q = 2.3h \frac{u_*}{\kappa} \log \frac{12h}{k_s} \qquad (7.12)$$

where h is the flow depth. If we accept that equation (7.4) from the Shields diagram is correct for the threshold of uniform coarse grains, then $u_* = [0.045(\rho_s - \rho)gD/\rho]^{1/2}$, and the depth at the threshold condition is given by the DuBoys relationship as $h_t = u_{*t}^2/gS$, where S is the channel slope. Making these substitutions in equation (7.12) yields the threshold discharge

$$q_t = 0.055 \left[\frac{\rho_s - \rho}{\rho}\right]^{3/2} g^{1/2} \frac{D^{3/2}}{S} \log \left[\frac{0.54(\rho_s - \rho)}{\rho \zeta S}\right] \quad (7.13)$$

where $k_s = \zeta D$ has been used to relate the roughness parameter to the grain size (the proportionality coefficient will be on the order of $\zeta \approx 1$ to 3 for a flat bed of uniform grains). Equation (7.13) predicts that $q_t \propto D^{3/2}$, which is seen to be the same as the proportionality of equation (7.7) established by Bathurst. The dependence on the slope S is more complex than found empirically by Bathurst in that equation (7.13) contains the slope within the logarithm term. It can be shown that

$$\log [0.54(\rho_s - \rho)/\rho \zeta S] \approx [0.54(\rho_s - \rho)/\rho \zeta S]^c,$$

where c ranges from 0.22 to 0.38 with an average 0.28 for the range of slopes $S = 0.01$ to 0.10 of interest. Hence the slope dependence in equation (7.13) is approximately equivalent to $q_t \propto S^{-1.28}$, which is on the same order as $q_t \propto S^{-1.12}$ found empirically by Bathurst in equation (7.7). Bettess (1984) has derived a relationship that is basically the same as equation (7.13), his derivation having been based on the friction-coefficient equation for fully rough beds. Bettess goes on to demonstrate that the resulting equation agrees with the same data that Bathurst *et al* (1983) employed to establish an earlier version of equation (7.7). These comparisons suggest the compatibility of defining the critical entrainment condition in terms of the unit-width discharge as well as mean bed stresses or the Shields entrainment function, at least for the uniform-bed case.

We can expand the comparison to selective entrainment by repeating the above derivation but using equation (7.6) for the entrainment stress rather than equation (7.4), which applies only to uniformly sized grains. This derivation yields

$$q_{ti} = 0.055 \left[\frac{\rho_s - \rho}{\rho}\right]^{3/2} g^{1/2} \frac{D_{50}^{0.9} D_i^{0.6}}{S} \log \left[\frac{0.54(\rho_s - \rho)}{\rho \zeta S} \frac{D_{bi}^{0.4}}{D_{50}^{0.4}}\right]. \quad (7.14)$$

Conversion of the logarithm term as before with $c = 0.28$ gives $q_{ti} \propto D_{50}^{0.8} D_{bi}^{0.7}$; the D_{bi} exponent is larger than those found by Bathurst (1987) in natural rivers (range 0.2 to 0.4) but less than the 1.22 exponent found in Figure 7.5 in the comparison with the data of Milhous (1973). It is possible that this range of exponents results from different sorting coefficients of bed materials or other aspects of bed-particle geometry, so these differences should not be taken as an indication of the incompatibility of the approaches. Although there are differences, there is still a semblance of agreement, and it is likely that the differences will be resolved by additional investigations. The significance is that

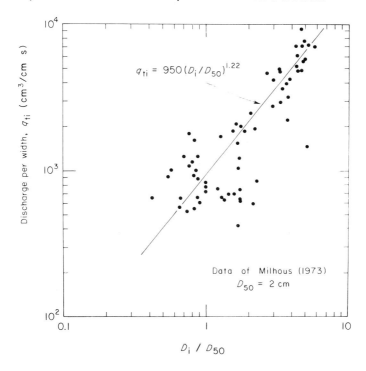

Figure 7.5 The selective-entrainment measurements of Milhous (1973) from a gravel-bed stream, used to establish a relationship between the entrainment discharge per unit flow width, q_{bi}, and D_i/D_{50} according to equation (7.9) from Bathurst (1987).

Bathurst's discharge relationships are apparently compatible with those based on mean bed stresses or the Shields entrainment function. Both approaches can be applied to flow-competence evaluations, even to the same deposit, to yield estimates of mean bed stresses and discharges.

THEORETICAL GRAIN-PIVOTING AND SLIDING EVALUATIONS

The above analyses involved correlations which yield empirical formulae that can be used in flow-competence evaluations. However, they have contributed little to our understanding of the processes of selective grain entrainment and are a poor basis for future advancements. The processes themselves can be examined through the development of models which consider the forces of fluid flow and particle resistance. The earliest analyses of this type were those by Airy (1885) and Law (1885) who respectively considered the forces of a current required to roll or slide a cube-shaped particle on a flat stream bed. More modern considerations of such analyses began with the studies of Rubey (1938) and

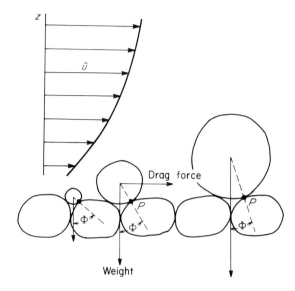

Figure 7.6 Grains in pivoting position for entrainment, illustrating that the larger the grain the greater its exposure to the flow and the smaller its pivoting angle, Φ.

White (1940), and recent examples are those of Slingerland (1977), Phillips (1980), Naden (1987) and Komar (1988).

Important to the application of such models to selective-entrainment processes are considerations of size-dependent factors such as a grain's exposure to the flow and its pivoting angle over the underlying particles (Slingerland, 1977). It is seen in Figure 7.6 that a larger-than-average grain not only projects further up into the flow, and hence experiences more drag, but it also has the smallest pivoting angle, Φ, which will allow this particle to rotate more easily out of its resting position. To be successful, any model of selective grain entrainment must account for such effects.

For a large clast resting on a bed of generally smaller grains, the condition of interest to competence evaluations, the balance of the forces at the critical instant of entrainment can be analyzed in terms of the velocity $\bar{u}_p$ measured at the level of the grain's center. A detailed derivation is presented in Komar (1988), an analysis which yields

$$\bar{u}_{pi} = (\Omega/A) \left[\frac{4}{3} \frac{\rho_s - \rho}{\rho} g D_{bi} \frac{\tan \Phi}{1 + (B/A)(D_{bi}/D_{ci}) \tan \Phi} \right]^{1/2}, \quad (7.15)$$

where D_{bi} and D_{ci} are respectively the grain's intermediate and shortest axial diameters (it has been assumed that the clast is oriented with its longest axial diameter across the flow width, the intermediate diameter in the flow direction, and the shortest diameter vertical). The i subscript is again used as a reminder that this is a selective-entrainment relationship for the movement of an individual particle from a deposit of mixed sizes. The A and B coefficients incorporate parameters important to the lift and drag forces acting on the particle; B/A is approximately equal to the ratio of lift to drag. The Ω proportionality coefficient converts the instantaneous velocity, which actually causes entrainment, into the $\bar{u}_{pi}$ time-averaged velocity that can be related to the discharge and other flow and channel parameters. Ranges of values for these three coefficients are poorly established, so that only approximate magnitudes can be assigned.

Equation (7.15) might be suitable for competence calculations, especially those involving erratics, where there is an isolated cobble or boulder such that the $\bar{u}_{pi}$ velocity is at a reasonable distance above the average bed level. However, in general this velocity would be difficult to measure, especially for grains that are contained within the general distribution of bed-sediment sizes. An additional problem with equation (7.15) is that $\bar{u}_{pi}$ is not at a fixed distance above the bed, both the distance and $\bar{u}_{pi}$ increasing with grain size. This has the effect of obscuring the role of grain exposure in the selective entrainment of different sizes.

The relationship between $\bar{u}_{pi}$ and u_* (and hence τ) can be determined from the velocity–profile relationship of equation (7.11), yielding

$$\bar{u}_{pi} = 2.3 \frac{u_*}{\kappa} \log \frac{30 z_{pi}}{\zeta D_{50}}, \qquad (7.16)$$

having again assumed that the bottom roughness is proportional to the median diameter of the bed material ($k_s = \zeta D_{50}$). Equation (7.15) can then be converted to

$$\theta_{ti} = \frac{(4/3)(\Omega^2/A)}{[5.75 \log(30 z_{pi}/\zeta D_{50})]^2} \left[\frac{\tan \Phi}{1 + (B/A)(D_{bi}/D_{ci}) \tan \Phi} \right]. \qquad (7.17)$$

There are two parts to this relationship which separately account for selective entrainment due to grain exposure and pivoting-angle variations. The grain-exposure effect enters via the dependence on z_{pi}/D_{50}, the z_{pi} distance of the grain's center above the bed compared with the median diameter of the bed material. As expected, according to equation (7.17) the entrainment θ_{ti} of a particle will decrease with increasing z_{pi}/D_{50}, that is, with increasing projection above the bed for a given particle weight. In the derivation z_{pi} is defined as the distance above the $u = 0$ level of the velocity profile, but in most applications that level will not be known with certainty. In the case of erratics the value of z_{pi} will be sufficiently large that a precise measure with respect to the $u = 0$ level will not be required in order to yield reasonable results; in this case $z_{pi} \approx 0.5 D_{ci}$ if the grain is oriented with its smallest axial diameter vertical with respect to the bed.

The bracketed term of equation (7.17) contains the dependence of θ_{ti} on the grain's pivoting angle, Φ. As noted in Figure 7.6, from the geometry of the particles it can be seen that Φ will decrease as the size of the particle increases relative to the underlying grains. The studies of Miller and Byrne (1966) and Li and Komar (1986) have examined such variations, and their measurements support the empirical relationship

$$\Phi = e(D_{bi}/D_{50})^{-f} \qquad (7.18)$$

given in a form that is required here. The e and f coefficients are empirical, and Miller and Byrne and Li and Komar provide values for spheres, ellipsoidal grains with and without imbrication, and for angular gravel. Komar and Li (1986) have shown that these variations in Φ can be important to selective entrainment by grain size and shape. The pivoting angle enters both the numerator and denominator of the bracketed term of equation (7.17) owing to having included fluid lift forces as well as drag. However, the dependence is such that an increase in D_{bi}/D_{50} produces a decrease in Φ according to equation (7.18), and this in turn produces a decrease in θ_{ti} as expected. Thus, selective entrainment depends on the two ratios $z_{pi}/D_{50} \approx 0.5 D_{ci}/D_{50}$ and D_{bi}/D_{50}, the first accounting for the particle's exposure due to its projection above the general bed level while the latter determines variations in pivoting angles. The size and density of the particle itself are contained within $\theta_{ti} = \tau_{ti}/(\rho_s - \rho)g D_{bi}$ and emerge when the τ_{ti} entrainment stress is calculated for a specific D_{bi} from the θ_{ti} value determined from equation (7.17).

The derivation which led to equations (7.15) and (7.17) is easily modified for the conditions of grain sliding and rolling. The forces of fluid drag and lift remain, the resisting force now being that of friction, which replaces pivoting; the resulting equations involve a simple substitution of a μ_f friction coefficient for tan Φ in the relationships. The modification of equation (7.15) yields a relationship that is similar to those developed by Airy (1885) and Mears (1979), and applied to flow-competence evaluations by Bradley and Mears (1980). Friction is important for both sliding and rolling, although the μ_f coefficients for sliding will be greater than those for rolling. Unfortunately, satisfactory data are nearly non-existent for either condition (see summary in Komar, 1988).

Applications of the above equations require values for the A, B and Ω coefficients. As discussed in Komar (1988), existing studies are a poor guide in that they have major disagreements on such important matters as the relative significance of lift and drag forces which determine the ratio B/A. In the application to coarse-grained entrainment and competence evaluations, it is necessary to assume a value for B/A with some guidance from that literature. However, once B/A has been selected, the choice of the Ω^2/A proportionality factor in equation (7.17) can be that which yields $\theta_{ti} = 0.045$ for the case in which $D_{bi}/D_{50} = 1$ (Komar, 1988). This selection again ensures agreement with the

Shields curve of Miller *et al.* (1977) for the special case of uniformly sized bed materials.

Adequate data for testing equations (7.16) and (7.17) are presently unavailable. Laboratory measurements such as those of Fenton and Abbott (1977) have demonstrated the important role of grain exposure in selective entrainment, in effect having shown that θ_{ti} does decrease with increasing z_{pi}/D_{50}. However, it is apparent from the nature of their experiments that there must have been a parallel role played by pivoting angle variations, the pivoting angle decreasing as they moved a grain upward so as to increase its exposure; unfortunately this dependence was not accounted for in the study of Fenton and Abbott.

Of more interest here is to see how well equation (7.17) predicts the data trends found by studies such as those of Milhous (1973), Carling (1983) and Hammond *et al.* (1984) since this would provide a better indication of the potential applicability to field evaluations of flow competence. Such a comparison is shown in Figure 7.7 from Komar (1988). A series of curves is given, curve (c) being for the case of grain sliding where all of the sorting is due to relative projection distances, and curve (d) is the other extreme where all of the sorting is attributed to variations in the pivoting angles. The most reasonable results are obtained with curve (b) which includes both grain exposure and

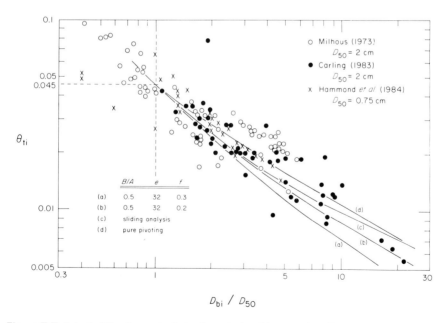

Figure 7.7 A test of the theoretical pivoting relationship, equation (7.17), and a comparable formula for entrainment by particle sliding, compared with the data for the entrainment of gravel (From Komar, 1988).

pivoting effects, and was computed with equation (7.17) with a lift-to-drag ratio $B/A = 0.5$, coefficients $e = 32$ and $f = 0.2$ in the pivoting-angle relationship of equation (7.18), and with $\Omega^2/A = 2.89$ so as to yield $\theta_{ti} = 0.045$ when $D_{bi}/D_{50} = 1$. These coefficients are acceptable, and the good agreement with the combined data demonstrates that process models of grain entrainment by pivoting can serve as the basis for further advances and improved predictions of selective entrainment and flow competence.

OTHER FACTORS INVOLVED IN COMPETENCE EVALUATIONS

There are other factors that affect flow-competence evaluations which involve corrections or modification of the results from the above formulae. In competence analyses we are commonly dealing with mountain streams having steep slopes and containing large cobbles or boulders that are on the order of the flow depth. In order to obtain reasonable results we must be able to make quantitative corrections for such factors.

Slope effects can be accounted for in the grain-pivoting and sliding models by including the downstream component of the particle's weight. This predicts that for the case of grain pivoting the correction will be

$$\theta_t'' = \theta_t [\cos ß(\tan \Phi - \tan ß)]^{-1}, \qquad (7.19)$$

where θ_t is the uncorrected Shields entrainment function dealt with in the earlier analyses, while θ_t'' is the corrected value which accounts for the channel slope angle ß. The corrections for entrainment stresses and discharges take the same form as this relationship. For uniform grains $\tan \Phi \approx 0.6$, so that this correction does not become significant until channel slopes of about $\tan ß > 0.1$. However, it has been seen that Φ decreases for large particles pivoting over smaller grains (Figure 7.6), and according to equation (7.18) $\tan \Phi$ could be as low as 0.2, so that slope corrections are potentially more important in selective entrainment and competence evaluations. Unfortunately, there has not been an adequate test of equation (7.19) when applied to large grains moving over smaller grains. The relationship has been verified by Luque and Van Beek (1976) for beds of uniform sand in flume experiments with slopes up to 22° ($\tan ß = 0.40$). This verification together with its basic derivation indicates the validity of a slope correction according to equation (7.19), but attempts should still be undertaken to test it for the entrainment of coarse-grained materials.

One problem with confirming this slope correction in gravel-bed streams is the parallel occurrence of the effects of low ratios of flow depths to gravel diameters. The importance of this h/D_{50} ratio is illustrated by the flume experiments of Ashida and Bayazit (1973). Channel slopes ranged from 0.01 to 0.20, but an attempt was made to remove the slope effect by employing equation (7.19). The

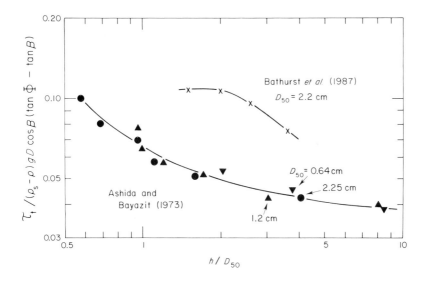

Figure 7.8 The data of Ashida and Bayazit (1973) from flume experiments which demonstrate that the Shields entrainment function increases at low ratios of flow depth to bed roughness.

resulting corrected θ_t'' values are plotted versus h/D_{50} in Figure 7.8 where it is seen that as h/D_{50} increases, θ_t'' asymptotically approaches the expected 0.045 value established by the Shields curve (Miller et al., 1977). The effects of low depth-to-diameter ratios do not become appreciable until $h/D_{50} < 3$, beyond which corrections of estimated entrainment θ_t, stresses and discharges should be attempted. The results from Ashida and Bayazit indicate that the corrections could be significant, involving a factor of 2 where $h/D_{50} < 1$, that is, when the particles actually project above the water's surface. Bathurst et al. (1983, 1987) obtained additional measurements of variations of θ_t'' with h/D_{50}, but the data scatter was large. Bettess (1984) re-analyzed the measurements and concluded that the scatter can be reduced if the analysis is in terms of h/k_s rather than h/D_{50}, and where this k_s roughness length is calculated from the rough-turbulent friction coefficient law. Further investigations are warranted, but present studies can serve as the basis for corrections in competence evaluations.

The trend of the θ_t increase due to low h/D_{50} or h/k_s ratios is illustrated by the short-dashed lines in Figure 7.9, the trends one obtains by plotting the measurements of Ashida and Bayazit (1973) on the Shields diagram. Also graphed are the downward sloping lines established by the selective-entrainment measurements of Hammond et al. (1984) and of Carling (1983). Carling's data from Great Eggleshope Beck plot in the expected position, crossing the standard Shields curve at the D_{50} of the stream gravels. However, his data from Carl Beck establish a line that slopes downward as expected, but is displaced well above the

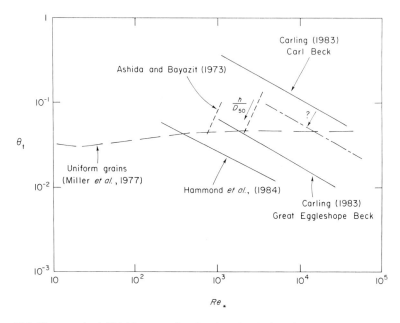

Figure 7.9 The standard Shields curve for the threshold of uniform grains (long-dashed curve), compared with the selective-entrainment curves based on the measurements of Carling (1983) and Hammond et al. (1984). The upward shift of the curve for Carl Beck might be accounted for by low h/D_{50} ratios as illustrated by the curves from Ashida and Bayazit (1973), although the shift could also be produced by the low ratio of channel width to depth of that stream.

Shields curve (Figure 7.9). Its 'correct' plotting position is indicated, based on $D_{50} = 7$ cm (Carling, Figure 7.3) and the assumption that these data from Carl Beck should be consistent with the other data sets. This vertical displacement above its expected plotting position might result from low h/D_{50} ratios in Carl Beck, the data moving upward to higher θ_t levels in the sense indicated by the dashed curves of Ashida and Bayazit (1973). However, there is still another possible factor, the low channel width-to-depth ratio of Carl Beck (Carling, 1983). The entrainment stresses reported by Carling are total flow stresses, the stress exerted on the bed which actually produces particle entrainment and transport, plus the drag stress acting on the banks of the stream. A significant drag on the banks could also displace the data to higher θ_t levels as seen for the data from Carl Beck. In contrast, Great Eggleshope Beck has a much higher width-to-depth ratio, so that total stresses and effective bed stresses are essentially the same. The importance of establishing the distribution of flow stresses and the portion actually acting on the bed is also illustrated by the analysis of Bathurst (1987). He demonstrated that q_t, the stress per unit flow width, should be based on only the portion of the channel width where there is active bed-load movement. Such corrections may be especially important in competence calculations where flood waters spill beyond the main channel.

Bathurst (1987) attempted to establish a correlation between the empirical exponent k of equation (7.9) and the D_{84}/D_{16} sorting coefficient of bed materials. Such a dependence is intuitively expected, and the b exponent in equation (7.5) for the entrainment θ_{ti} is likely also to be a function of sorting. However, these coefficients probably depend on the bed microtopography in addition to an overall sorting coefficient. Studies such as those of Brayshaw (1985), Laronne and Carson (1976) and Brayshaw et al. (1983) have demonstrated the importance of particle clusters and grain fabric in the transport of gravel. It remains to undertake investigations which document both the geometry of the bed particles and the resulting patterns of selective entrainment and transport. The hope is that such studies would explain the observed variations in the coefficients of the empirical selective-entrainment relationships, and ultimately lead to improved predictive capabilities. However, it is likely that the theoretical models of grain pivoting and sliding, equations (7.15) and (7.17), will provide a better basis for analyses of such effects in that those models can focus directly on factors such as grain imbrication and sheltering.

Related to the problems of dealing with bed morphology and grain packing is the necessity for selecting a representative D_{50} in the entrainment relationships, it having been established that the threshold of a large clast depends on D_{50} as well as on its own diameter. Gravel-bed streams typically display considerable sorting patterns with segregations of clasts according to size, so it is difficult in such a circumstance to decide upon a representative D_{50} value. This can be done as demonstrated by the studies of Carling (1983, 1987a, 1987b) in Great Eggleshope Beck. In our early analyses of that data (Komar and Li, 1986), we noted that the trend of those selective entrainment measurements cross the Shields curve at a point which corresponds to a 2 cm grain diameter, and accordingly took that size as D_{50} in the analyses. Unknown to us, Carling (1987b) had independently arrived at the same D_{50} value from an extensive sampling program (see discussion in Carling, 1987a). This problem of selecting D_{50} is also illustrated by analyses of the data of Milhous (1973) obtained in a stream where the bed is armored. If one takes $D_{50} = 2$ cm from Milhous's size analyses of the complete deposit, then the proportionality coefficient in equation (7.5) is $a = 0.045$, the expected value for the cross-over with the Shields curve and the value found for other data sets. If the analysis had instead been based on $D_{50} = 6$ cm, the median of the armor surface material, then the proportionality coefficient would have been $a = 0.027$. Therefore, this coefficient is sensitive to the selection of D_{50} in the data analysis. In the same way, the choice of D_{50} to be used in equations (7.6) and (7.9) in a competence evaluation greatly affects the resulting values of the flow stress and discharge.

EXAMPLES OF FLOW-COMPETENCE EVALUATIONS

In this section a series of examples will be presented which illustrate applications of the flow-competence equations. These will progress from modern rivers where reasonable confirmation of the results can be obtained, to catastrophic floods produced by glacial-lake outbursts, and finally to gravel deposits in the rock record where there is little opportunity to test the validity of the calculated hydraulic parameters.

The first 'application' will examine the flow conditions required for entrainment of the largest clasts in Carling's (1983) tagging experiments in Great Eggleshope Beck. In that these data have been used to establish equations (7.5) and (7.6), this example cannot serve as a true test of those specific relationships. However, the computations will still serve to illustrate the degree of potential errors involved in utilizing relationships such as equations (7.1) and (7.4), relationships that have been used in many applications.

The largest clast used by Carling (1983) in his tagging experiments had an intermediate axial diameter of 44 cm, and it was determined by monitoring the movement of this boulder that a stress of 404 dynes/cm^2 is required for its entrainment. Equation (7.6) with D_{50} = 2 cm and D_{bi} = 44 cm yields τ_{ti} = 443 dynes/cm^2, in good agreement with the measured stress as expected, since the measurements of Carling were included in the data used to establish that relationship. Equation (7.4) based on flume experiments with uniform grains predicts a threshold stress τ_t = 3200 dynes/cm^2 for that 44 cm diameter clast, a factor 8 larger than the measured stress. The competence equation (7.1) from Costa (1983) predicts a flow stess τ_c = 2590 dynes/cm^2 for that diameter, again greatly in excess of the 404 dynes/cm^2 measured stress. Similar results are obtained in comparisons with other data sets from modern rivers. This is apparent in Figure 7.2 since the coarse-grained ends of the selective-entrainment curves for the various data sets are all below the line from equation (7.4) for the threshold of uniform grains and the competence line from equation (7.1). It is further apparent that the larger the magnitude of the flood and the larger the clast-size moved, the greater the resulting discrepancy (assuming that the selective-entrainment lines can be extrapolated beyond their present empirical limits). In view of such comparisons, I have recommended that flow-competence relationships such as equation (7.1) no longer be employed (Komar, 1987a), and it is also apparent that equation (7.4) for the threshold of uniform grains would yield unacceptable results. Although this 'application' has been to a modern stream, similar errors would be expected in applications of equations (7.1) and (7.4) to competence calculations for ancient deposits.

Kehew (1982), Kehew and Lord (1986) and Lord and Kehew (1987) have documented the morphology of spillways eroded by the catastrophic drainage of Glacial Lake Regina in Saskatchewan, Canada, and have analyzed the paleohydrology of the flood. Their studies provide measurements of the transported sediment sizes which permit flow-competence calculations, and their determinations of channel dimensions allow at least partial testing of the results.

Most of the focus of the analyses of Lord and Kehew (1987) was on the Elcott Bar in the Souris Spillway leading from Lake Regina. The largest boulder has a diameter of 154 cm, and the median diameter of the deposit determined from multiple size distributions is given by Lord and Kehew (Table 1) as D_{50} = 3 cm. This yields D_{bi}/D_{50} = 51, which goes somewhat beyond the data base used to establish equation (7.5), but if we accept some extrapolation that relationship yields $\theta_{ti} \approx 0.0035$ (having used $a = 0.045$ and $b = -0.65$ as established in Figure 7.3). The corresponding mean bed stress is τ_{ti} = 910 dynes/cm^2 (using ρ_s = 2.7 g/cm^3 for the grain density as measured by Lord and Kehew). The slope of the Souris Spillway is S = 0.00019, and according to the DuBoys equation, the flow depth would have been $h = \tau_{ti}/\rho g S \approx$ 48 m according to these flow-competence calculations. This value is effectively the same as the 45 m depth established by Lord and Kehew in their field measurements of the eroded channel at the Elcott Bar location. We therefore have reasonable confirmation of the results.

Lord and Kehew (1987) attempted to use competence relationships such as equation (7.1), but concluded that the results are unreasonable. With D = 154 cm for the maximum boulder found on Elcott Bar, equation (7.1) predicts a competence flown stress $\tau_c = 1.2 \times 10^4$ dynes/cm^3, a factor 13 greater than found above with equation (7.5). Use of the DuBoys equation yields a corresponding flow depth $h \approx 630$ m, clearly an unacceptable result, just as concluded by Lord and Kehew. Equation (7.4) for uniform grains yields even greater stresses and flow depths.

The Bathurst equation (7.9) yields a discharge per unit width $q_{ti} = 1.17 \times 10^6$ cm^3/(cm s) = 117 m^3/(m s) for the D_{bi}/D_{50} = 51 ratio of the 154 cm boulder on Elcott Bar [$q_t = 3.6 \times 10^5$ cm^3/cm s from equation (7.7) for D_{50} = 3 cm, and taking k = 0.3 from Bathurst (1987, Figure 7.3) for the exponent in equation (7.9)]. For the 1 km channel width at Elcott Bar (Lord and Kehew, 1987), the predicted discharge is 1.17×10^5 m^3/s, and the mean velocity would have been u = 2.6 m/s for a 45 m depth. Both values are on the same order as those calculated by Kehew (1982) and Lord and Kehew (1987) using the Manning equation. For agreement between this 2.6 m/s velocity and the τ_{ti} = 910 dynes/cm^2 stress calculated with equations (7.5) or (7.6), the Darcy–Weisbach friction coefficient would have to be f = 0.11 according to the relationship $u = u_* \sqrt{(8/f)}$, where $u_* = \sqrt{(\tau/\rho)}$ is the shear velocity. This value for f is reasonable, indicating that the separate calculations of τ_{ti} and q_{ti} by the two approaches are compatible.

As an example of flow-competence evaluations for gravels found in the geologic rock record, the techniques will be applied to the turbidite deposits which occur in the late Miocene lower Capistrano Formation of California near Dana Point. This will give me the opportunity to correct the calculations presented in an earlier publication (Komar, 1970). The outcrop was described by

Bartow (1966) who interpreted the lithologic relationships as that of a cross-section of a Miocene submarine fan channel which he named Doheny Channel. The channel fill consists of massive coarse-grained pebbly sandstone, and a conglomerate layer which may have been the thalweg of the channel. Clasts of 20 cm diameter are reasonably common in this conglomerate, and it was the entrainment of that size I analyzed in my 1970 paper. Unfortunately, I used equation (7.4) for the threshold of uniform grains according to the Shields curve, which we now see yields unacceptable results for the transport of coarser-than-average clasts within a mixture. Having used a 0.06 coefficient rather than 0.045 as given here in equation (7.4), my results were still more in error. If we revise those calculations but still use equation (7.4), then the estimate yields a flow stress τ_t = 1400 dynes/cm^2 and velocity 605 cm/s for movement of the 20 cm cobbles by a turbidity current having a density of 1.10 g/cm^3. However, the photographs suggest a median diameter of about D_{50} = 3 cm, giving D_{bi}/D_{50} = 20/3 = 6.7, and this yields θ_{ti} = 0.013 with equation (7.5) (a = 0.045 and b = −0.65) and τ_{ti} = 411 dynes/cm^2 with equation (7.6); the corresponding velocity is now only 327 cm/s, about half that determined with equation (7.4).

It is not generally possible to verify the results in such applications of flow-competence evaluations for gravels found in the geologic rock record. In this specific example, evidence from turbidity currents in the present deep sea indicates that flows can achieve velocities up to 20 m/s, and 10 m/s may be common. Therefore, the 3.3 m/s velocity calculated above for the Miocene Doheny Channel is quite modest and well within the range of reasonable values.

SUMMARY OF CONCLUSIONS

The objective of this report has been to provide an overall assessment of flow-competence evaluations. It has been seen that, thanks to recent studies of selective entrainment of gravels from deposits of mixed sizes, competence evaluations can now better account for such sorting processes. Comparisons of those data with formulae such as equation (7.1), which have been used in past competence assessments, demonstrate their inadequacy and tendency to significantly over-estimate flood-flow stresses and discharges. Competence evaluations should instead be based on selective-entrainment relationships such as equations (7.5) and (7.6) which yield the mean-flow stress, and on equation (7.9) which gives the discharge per unit flow width. Once those estimates have been made, corrections or refinements are possible which account for steep channel slopes and low ratios of flow depths to bed roughness. Therefore, if flow competence is recognized as a process of selective entrainment, then the technique is valid and it is possible to make reasonable assessments of flood hydraulics. However, in all cases attempts should be made at independent confirmations of the results.

The present review has also indicated a number of deficiencies in our understanding which require further investigation. In my view, a few of the more important are:

(1) Although the entrainment relationships which respectively yield mean-flow stresses and discharges appear to be basically compatible, more study is needed to establish their agreement.
(2) Additional data are needed to establish how the coefficients in the empirical relationships depend on grain sorting, bed packing, etc., especially the b and k exponents respectively in equations (7.5) and (7.9).
(3) How is one to choose a representative D_{50} median diameter to be used in the selective-entrainment relationships when faced with deposits that are affected by horizontal sorting and vertical armoring?
(4) What is the relative importance of grain pivoting versus sliding, and how can we better focus on these processes so as to test and improve the theoretical relationships such as equations (7.15) and (7.17)?
(5) How can we improve our quantitative assessments of the entrainment of erratics, particles which are individually much larger than the general bed sediments? What are the conditions which lead to scour and their burial rather than downstream transport?

The elimination of such deficiencies will greatly improve our capacity for making flow-competence calculations.

ACKNOWLEDGEMENTS

This review was undertaken with support from the Planetary Geology and Geophysics Program, Solar System Exploration Division, NASA. I would like to thank James Bathurst and Zhenlin Li for their helpful suggestions in reviewing this paper.

REFERENCES

Airy, W. (1885). Discussion on non-tidal rivers, *Proc. Institution of Civil Engrs.*, **82**, 25–6.
Andrews, E. D. (1983). Entrainment of gravel from naturally sorted riverbed material, *Geol. Soc. Amer. Bull.*, **94**, 1225–31.
Andrews, E. D., and Erman, D. C. (1986). Persistence in the size distribution of surficial bed material during an extreme snowmelt flood, *Water Resources Research*, **22**(2), 191–7.
Ashida, K., and Bayazit, M. (1973). Initiation of motion and roughness of flows in steep channels, *International Association for Hydraulic Res., Proc. 15th Congress*, **1**, 475–84.

Baker, V. R. (1973). Paleohydrology and sedimentology of Lake Missoula flooding in eastern Washington, *Geol. Soc. Amer. Special Paper*, **14**, 79 pp.

Baker, V. R., and Ritter, D. F. (1975). Competence of rivers to transport coarse bedload material, *Geol. Soc. Amer. Bull.*, **86**, 975–8.

Bartow, J. A. (1966). Deep submarine channel in the upper Miocene, Orange County, California, *Jour. Sed. Petrology*, **36**, 700–5.

Bathurst, J. C. (1987). Critical conditions for bed material movement in steep, boulder-bed streams, in R. L. Beschta *et al.*, (eds), *Erosion and Sedimentation in the Pacific Rim*, IAHS Publ. No. 165, pp. 309–18.

Bathurst, J. C., Graf, W. H., and Cao, H. H. (1983). Initiation of sediment transport in steep channels with coarse bed material, in B. Multlu Summer and A. Muller (eds), *Mechanics of Sediment Transport*, Proc. Euromech 156, A. A. Balkema, Rotterdam, pp. 207–13.

Bathurst, J. C., Graf, W. H., and Cao, H. H. (1987). Bed load discharge equations for steep mountain rivers, in C. R. Thorne *et al.*, (eds), *Sediment Transport in Gravel-Bed Rivers*, Wiley, pp. 453–77.

Bettess, R. (1984). Initiation of sediment transport in gravel streams, *Proc. Institution of Civil Engineers*, Part 2, **77**, 79–88.

Bradley, W. C., and Mears, A. I. (1980). Calculations of flows needed to transport coarse fraction of Boulder Creek alluvium at Boulder, Colorado: Summary, *Geol. Soc. Amer. Bull.*, Part I, **91**, 135–8.

Brayshaw, A. C. (1985). Bed microtopography and entrainment thresholds in gravel-bed rivers, *Geol. Soc. Amer. Bull.*, **96**, 218–23.

Brayshaw, A. C., Frostick, L. E., and Reid, I. (1983). The hydrodynamics of particle clusters and sediment entrainment in coarse alluvial channels, *Sedimentology*, **30**, 137–43.

Carling, P. A. (1983). Threshold of coarse sediment transport in broad and narrow natural streams, *Earth Surface Processes*, **8**, 1–18.

Carling, P. A. (1987a). Discussion, *Sedimentology*, **34**, 957–9.

Carling, P. A. (1987b). Bed stability in gravel streams with reference to stream regulation and ecology, in K. S. Richards, (ed.), *Rivers*, Special Publication of the Transactions of the Institute of British Geographers, pp. 321–47.

Costa, J. E. (1983). Paleohydraulic reconstruction of flash-flood peaks from boulder deposits in the Colorado Front Range, Geol. Sco. Amer. Bull., **94**, 986–1004.

Day, T. J. (1980). A study of initial motion characteristics of particles in graded bed material, *Geol. Survey of Canada, Current Research*, Part A, Paper 80-1A, 281–6.

Fahnestock, R. K. (1963). Morphology and hydrology of a glacial stream—White River, Mount Rainier, Washington, *U.S. Geol. Survey Prof. Paper*, **422-A**, 1–70.

Fahnestock, R. K., and Haushild, W. L. (1962). Flume studies of the transport of pebbles and cobbles on a sand bed, *Geol. Soc. Amerc. Bull.*, **73**, 1431–6.

Fenton, J. D., and Abbott, J. E. (1977). Initial movement of grains on a stream bed: the effect of relative protrusion, *Proc. Royal Soc. London*, Series A, **352**, 523–7.

Gilbert, G. K., and Murphy, E. C. (1974). The transportation of debris by running water, *U.S. Geol. Survey Prof. Paper*, **86**, 263 p.

Hammond, F. D. C., Heathershaw, A. D., and Langhorne, D. N. (1984). A comparison between Shields' threshold criterion and the movement of loosely packed gravel in a tidal channel, *Sedimentology*, **31**, 51–62.

Helley, E. J. (1969). Field measurement of the initiation of large particle motion in Blue Creek near Klamath, California, *U.S. Geol. Survey Prof. Paper*, **562-G**, 19 p.

Kehew, A. E. (1982). Catastrophic flood hypothesis for the origin of the Souris spillway, Saskatchewan and North Dakota, *Geol. Soc. Amer. Bull.*, **93**, 1951–8.

Kehew, A. E., and Lord, M. L. (1986). Origin of large-scale erosional feathers of glacial-lake spillways in the northern Great Plains, *Geol. Soc. Amer. Bull.,* **97**, 162–77.

Komar, P. D. (1970). The competence of turbidity current flow, *Geol. Soc. Amer. Bull.,* **81**, 1555–62.

Komar, P. D. (1987a). Selective gravel entrainment and the empirical evaluation of flow competence, *Sedimentology,* **34**, 1165–76.

Komar, P. D. (1987b). Selective grain entrainment by a current from a bed of mixed sizes: A reanalysis, *Jour. Sedimentary Petrology,* **57**, 203–11.

Komar, P. D. (1988). Applications of grain-pivoting and sliding analyses to selective entrainment and flow-competence evaluations, *Sedimentology,* **35**, 681–695.

Komar, P. D., and Li, Z. (1986). Pivoting analyses of the selective entrainment of sediments by shape and size with application to gravel threshold, *Sedimentology,* **33**, 425–36.

Krumbein, W. C., and Lieblein, J. (1956). Geological application of extreme-value methods to interpretation of cobbles and boulders in gravel deposits, *Amer. Geophys. Union Trans.,* **37**, 313–19.

Laronne, J. B., and Carson, M. A. (1976). Interrelationships between bed morphology and bed-material transport for a small, gravel-bed channel, *Sedimentology,* **23**, 67–85.

Law, H. (1885). Discussion on non-tidal rivers, *Proc. Institution of Civil Engrs.,* **82**, 29–31.

Li, Z., and Komar, P. D. (1986). Laboratory measurements of pivoting angles for applications to selective entrainment of gravel in a current, *Sedimentology,* **33**, 413–23.

Lord, M. L., and Kehew, A. E. (1987). Sedimentology and paleohydrology of glacial-lake outburst deposits in southeastern Saskatchewan and northwestern North Dakota, *Geol. Soc. Amer. Bull.,* **99**, 663–73.

Luque, R. F., and Van Beek, R. (1976). Erosion and transport of bed-load sediment, *Journal of Hydraulic Research,* **14**, 127–44.

Mears, A. I. (1979). Flooding and sediment transport in a small alpine drainage basin in Colorado, *Geology,* **7**, 53–7.

Milhous, R. T. (1973). *Sediment Transport in a Gravel-bottomed Stream,* Unpublished Ph.D thesis, Oregon State Univ., Corvallis, 232 p.

Miller, M. C., McCave, I. N., and Komar, P. D. (1977). Threshold of sediment motion in undirectional currents, *Sedimentology,* **24**, 507–28.

Miller, R. L., and Byrne, R. J. (1966). The angle of repose for a single grain on a fixed rough bed, *Sedimentology,* **6**, 303–14.

Naden, P. (1987). An erosion criterion for gravel-bed rivers, *Earth Surface Processes and Landforms,* **12**, 83–93.

Novak, I. D. (1973). Predicting coarse sediment transport: The Hjulstrom curve revisited, in M. Morisawa, (ed.), *Fluvial Geomorphology,* Binghampton, New York, pp. 13–25.

Parker, G., Klingeman, P. C., and MacLean, D. G. (1982). Bedload and size distribution in paved gravel-bed streams, *J. Hydraulics Div., Amer. Soc. Civil Engrs.,* **108**(HY4), 544–71.

Phillips, M. (1980). A force balance for particle entrainment into a fluid stream, *J. Physics, D. Applied Physics,* **13**, 221–33.

Rubey, W. W. (1938). The forces required to move particles on a stream bed, *U.S. Geol. Survey Prof. Paper,* **189-E**, 121–41.

Slingerland, R. L. (1977). The effects of entrainment on the hydraulic equivalence relationships of light and heavy minerals in sands, *J. Sedimentary Petrology,* **47**, 753–70.

White, C. M. (1940). The equilibrium of grains on the bed of a stream, *Proc. Royal Soc. London,* Series A, **174**, 332–8.

Yalin, M. S., and Karahan, E. (1979). Inception of sediment transport, *J. Hydraulics Div., Amer. Soc. Civil Engrs.,* **195**(HY11), 1433–43.

… 135

8 Floods and Flood Sediments at River Confluences

IAN REID
Birkbeck College, London

JAMES L. BEST
University of Leeds, Leeds

LYNNE E. FROSTICK
Royal Holloway and Bedford New College, Egham

INTRODUCTION

Confluences are points of complex hydraulic adjustment in all drainage nets, whether natural or artificial. It is within the confluence precinct that two streams compete for limited space. Because of this, confluences are particularly prone to bank erosion (Mosley, 1975, 1976; Ashmore, 1982) and changing patterns of sedimentation (Krigström, 1962; Lodina and Chalov, 1971; Ashmore and Parker, 1983; Best, 1986). They are therefore of considerable interest to river engineers anxious to minimize the encroachment of rivers on riparian developments (Muskatirovic and Miloradov, 1980), and to maintain water-supply feeder canals (e.g. Behlke, 1964; Gildea and Wong, 1967; US Corps of Engineers, 1970).

Despite the important role of confluences, they have attracted little attention when compared with other elements of the drainage net, probably because they are spatially discrete and complex to model hydraulically. Nevertheless, there have been studies of flow (Taylor, 1944; Bowers, 1950; Behlke, 1964; Behlke and Pritchett, 1966; Itakura, 1972; McGuirk and Rodi, 1978; Best, 1985, 1987; Best and Reid, 1984, 1987), and the bedforms associated with river junctions (Krigström, 1962; Komura, 1973; Mosley, 1975, 1976, 1982; Ashmore, 1979, 1982; Ashmore and Parker, 1983; Kennedy, 1984; Best, 1985, 1986; Petts and Thoms, 1986, 1987); there have also been studies of the sedimentary consequences of combining one stream with another (Mosley and Schumm, 1977; Frostick and Reid, 1977, 1979; Knighton, 1980, 1982; Kochel and Baker, 1982; Alam *et al.*, 1985; Best, 1985, 1987; Best and Brayshaw, 1985). However, although work by Miller (1958), Richards (1980), Roy and Woldenberg (1986) and Roy and Roy (1988) has considered the changes in channel capacity at stream junctions, the hydraulic implications of combining the *flood waves* of two rivers rather than steady flows remains largely unexplored. Even flood routing studies have largely ignored these considerations although Prohaska (1978) does recognize the importance of flood wave coincidence at channel junctions in flood routing.

In fact, while recognizing that the *magnitude* of a flood wave generally increases downstream, an assumption is often made that the *shape* of that wave remains largely similar as it moves through a confluence. This is not unreasonable where tributary flows are small in comparison with the trunk stream. Neither has it seemed unreasonable to make this assumption in regions where rainfall is more or less uniformly distributed in time and space over a drainage net. However, there are a number of factors, both extrinsic as well as intrinsic to the drainage system, that are sufficiently influential to reduce the probability that the flood wave will remain essentially the same shape through a river confluence even in a perennial drainage system. In ephemeral systems that probability is reduced even further.

The fact that the flood hydrograph within a confluence can be highly variable has implications for sediment transport, for the generation of channel bedforms, and for bank erosion. This is the context for this study of the hydrology and sedimentology of river confluences.

CONFLUENCE LOCATION AND CHARACTER

Two confluences have been studied in detail. One is part of a perennial system and the other part of an ephemeral system, giving a wide range of flow regimes and their expected effects. The choice of confluences was governed by a number of factors that were peculiar to each study, such as the nature of the sediment carried by each stream and the junction angle. However, in both cases important prerequisites were that the combining streams had catchments and channels of similar size, and that each stream had an alluvial bed and was not confined by bedrock incision.

Perennial river confluence

The River Ure has a catchment of 50 km^2 and joins Widdale Beck (34 km^2) at a junction angle of 85° (Figure 8.1). Each drains an area that is underlain by Carboniferous limestones, sandstones and shales. Both valleys have been subject to Pleistocene glaciation, and as a result, are deepened and possess a considerable fill of morainic and fluvioglacial deposits (see e.g. Rose, 1980) which form a ready source of coarse alluvium for the present streams. The average rainfall of the area is 1916 mm/a, which is delivered throughout the year, mainly by frontal systems. The rain falls on peat moorlands, grazed

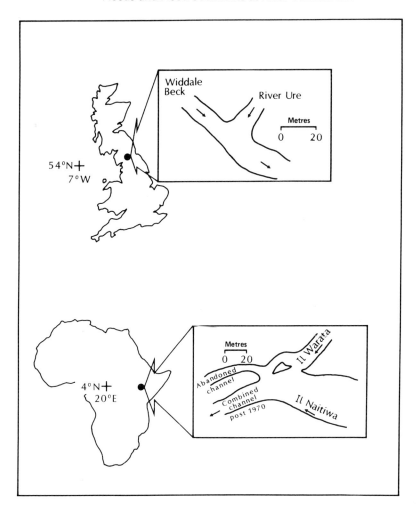

Figure 8.1 Location and planform of the Widdale Beck/River Ure and Il Warata/Il Naitiwa confluences.

permanent pasture and the improved grassland of the valley bottom. However, it is pertinent to note that the fraction of the Widdale Beck catchment that would be classed as 'valley bottom' and therefore under improved grass is much less than in the case of the River Ure. In both cases, rainfall-runoff response is reasonably rapid, and the time of concentration is typically 8.5 hours.

Just above the channel junction, the River Ure is 17 m wide, while Widdale Beck is 14 m wide. Each is incised into the common alluvial flood plain by 1–2 m. The channel slope of the Ure is 0.005, and that of Widdale Beck is 0.008. Each has a more or less planar pebble–cobble bed with a granule–sand matrix. Although the Ure catchment is larger, the Widdale Beck's flood regime as well as its higher channel slope appear to make it marginally more powerful. As a result, of the two, Widdale Beck acts as the trunk stream (Figure 8.1).

Water stage recorders were installed on both the River Ure and Widdale Beck about 450 m upstream of the junction, i.e. sufficiently far to avoid any backwater effects that might occur. The bed was repeatedly surveyed and marked particles were traced.

Ephemeral river confluence

The Il Warata drains a catchment of 24 km^2 which is largely underlain by Plio-Pleistocene basalts. It joins the Il Naitiwa whose catchment of 21 km^2 is largely on Pleistocene sediments (Figure 8.1). The combined river flows towards Lake Turkana in northern Kenya. The rainfall of the area averages 300 mm/a and is delivered by localised convective storms. The low rainfall and high potential evaporation of 3000 mm/a produce a sparsely vegetated semi-desert and ephemeral drainage. The nature of the rainfall means that two contiguous river catchments may have different runoff histories, with flows recorded in one and not in the other (Frostick *et al.*, 1983; Reid and Frostick, 1987).

Just above the confluence, the Il Warata is 25 m wide, while the Il Naitiwa is 40 m. Both are incised by 2–3 m into an alluvial plain. The site of the confluence has shifted upstream several times in recent years and has resulted in the Warata abandoning part of its former channel (Figure 8.1). The bed material of the Warata is a pebble–cobble mixture with granule–sand matrix; it is composed almost entirely of basalt. On the other hand, the Naitiwa bed is a sand–pebble–cobble mixture; clasts are composed of lithified sediments, basalts and tuffs that are mineralogically distinguishable from those carried by the Warata.

There are no discharge records for these two extremely remote ephemeral streams. However, by trenching the bed sediments the flood sequence has been interpreted by detailed examination of the channel fill.

DISCHARGE RELATIONS OF CONFLUENCING RIVERS

Gauging the flows of the River Ure and Widdale Beck at their confluence has allowed a number of patterns of discharge to emerge. Each has been differentiated either by the condition of the catchments prior to rainfall, i.e. a factor *intrinsic* to the drainage basin, or by the spatial distribution of rainfall, i.e. an *extrinsic* factor that carries with it an element of randomness.

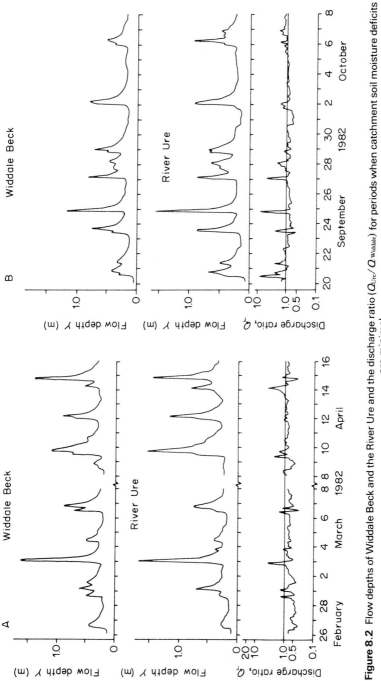

Figure 8.2 Flow depths of Widdale Beck and the River Ure and the discharge ratio ($Q_{Ure}/Q_{Widdale}$) for periods when catchment soil moisture deficits are minimal.

Coincident flood response of confluent rivers

There are periods of the year when flood waves on the River Ure and Widdale Beck are more or less coincident. These are times when soil moisture deficits are minimal and rainfall is uniformly distributed. In Britain, as with other regions in the temperate zones, this tends to occur during winter as a result of frontal rainfall. Figure 8.2 illustrates the pattern. The water stage of Widdale Beck and the River Ure are given for two selected parts of the record. Also given is the changing ratio of the River Ure discharge to that of Widdale Beck (Q_r). When examined in detail there are small but real differences in the rainfall-runoff relations of the two basins. For example, Widdale Beck appears to be more responsive to closely spaced rain storms as seen on the 28 February/1 March and 6 March. This may, in part, be attributed to within storm variability in rainfall intensity. However, in the main, the discharge ratio remains close to unity indicating that the responses of the catchments are sympathetic, and that the flood waves are more or less coincident.

Seasonal decline in tributary runoff

At a time of year when soil moisture deficits increase, it would be reasonable to expect a catchment to behave according to the amount of storage capacity created by evaporative losses. Such storage capacity will increase as the season lengthens. It will also vary between drainage basins, depending upon a host of variables including the moisture characteristics of local soils, topography, etc. Figure 8.3 shows a progressive change in the runoff response of the River Ure and Widdale Beck during that time of year when soil moisture deficits are at maximum. Towards the end of the summer, the Ure discharges less and less relative to Widdale Beck. An explanation for this response may lie in the fact that a higher fraction of the Ure catchment is bottomland, while its sideslopes are less steep than those of the Widdale. Because of this the Ure catchment moisture deficits are larger, storage is greater, and hydraulic conductivity of the soil is lower. However, in the Widdale basin, conductivity is higher so that soil water redistribution rates are higher and runoff generation is quicker. Additionally, the bottomland, wetted by interflow from the sideslopes, may still be sufficiently close to saturation to generate translatory flow. In either case, the runoff:rainfall ratio of Widdale Beck will fall off less quickly than that of the River Ure, producing the general decline in Q_r values (Figure 8.3).

Differences in basin concentration time

Even in situations where storm rainfall is evenly distributed, and where soil moisture deficits are minimal in the catchments of both rivers, there is ample scope for variable runoff response. In this case, the reasons for large short-term

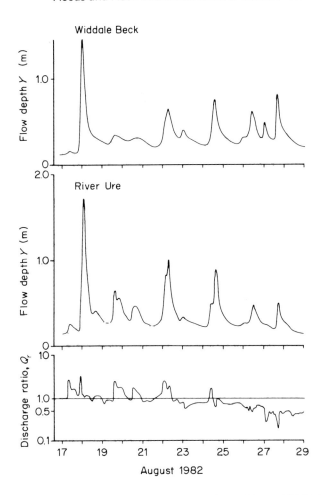

Figure 8.3 Flow depths and the discharge ratio of the River Ure and Widdale Beck for a period of increasing soil moisture deficit.

fluctuations in discharge ratio will be sought in the shape of the basins, in drainage densities (factors highlighted by the experimental studies of Black (1970), among others), in land use, and so on. Although a single cause cannot be identified, the storm of 2–3 April (Figure 8.4) amply illustrates the effect of differential basin runoff response on discharges entering the confluence. The Ure rises marginally more quickly, but falls in a simple recession curve. In contrast, Widdale Beck has a delayed but flashier rise and has a shoulder on the recession curve. As a result of these differences, the Ure temporarily dominates the confluence, is then relegated to second place by Widdale Beck, but partially recovers as its recession slows relative to that of the Beck. Similar trends in

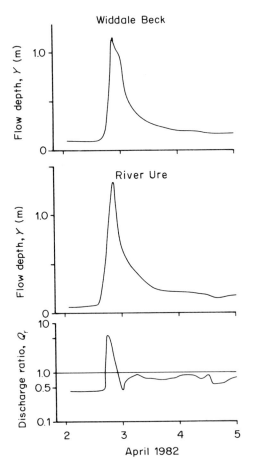

Figure 8.4 Flow depths and the discharge ratio of the River Ure and Widdale Beck showing the effects of differences in basin concentration time on the confluence flood hydrograph.

differential runoff response are evident in the February–April data of Figure 8.2 and are the chief cause of the fluctuations in discharge ratio at this time of year.

Differences in basin rainfall

Spatially discrete convective rain storms are characteristic of tropical and subtropical environments (e.g. Sharon, 1974), but are comparatively rare events in British weather. However, they do occur occasionally in summer. The effects of one such event were recorded at the Ure–Widdale confluence on 6–7 June 1982 (Figure 8.5). A localised storm delivered 33 mm to the Ure catchment while only 4 mm fell on that of the Widdale. A rise in water stage of 0.4 m on the Widdale

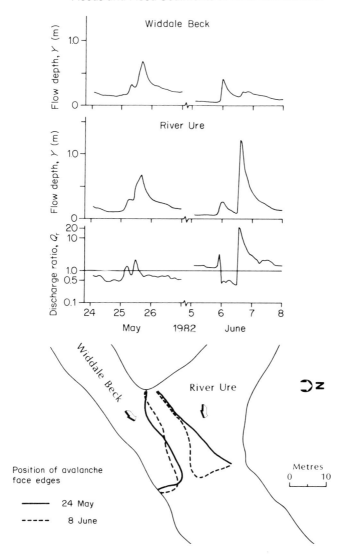

Figure 8.5 Flow depths and the discharge ratio of the River Ure and Widdale Beck showing the effect of a thunderstorm (6–7 June) in one drainage basin only (top). The effect of the thunderstorm of 6–7 June on bedform elements in the confluence precinct (bottom).

was overshadowed by a rise of 1.22 m in the Ure. The discharge ratio reached a maximum of 23 and, in contrast with the runoff pattern for evenly distributed rain, it remained >1 throughout the recession, reflecting continued domination of the confluence by the Ure on this occasion.

SEDIMENTARY CONSEQUENCES OF OUT-OF-PHASE FLOODS IN CONFLUENT RIVERS

Even though there are few quantitative studies (e.g. Best, 1985, 1987, 1988), the transport of sediment through river confluences is known to be complex. The pathways of individual bedload particles have been shown to depend upon the relative strength of flow in each stream (Best and Reid, 1987). However, the scour and bar forms of the whole confluence can be altered significantly for the same reason, with consequences for both flow and sediment transport during subsequent floods. The most dramatic changes are brought about when discharge ratios are very high, i.e. when the tributary flow dominates the confluence.

The Ure–Widdale flood of 6–7 June 1982 already discussed (Figure 8.5) is an example. The 3 m deep scour hole and the avalanche faces bounding it were mapped as part of a routine monitoring programme on 24 May. At that time, the avalanche faces were in their normal post-winter position indicating the marginally dominant part played by Widdale Beck towards the end of flood waves generated by evenly distributed rainfall. The confluence was mapped again after the 6–7 June flood and revealed a remarkable incursion of the Ure's bar into the scour hole (Figures 8.5 and 8.6). This field evidence complements flume observations by Ashmore (1979, 1982), Jaeggi (1986) and Best (1988) on the response of bed morphology to changing discharge ratio. Work by Petts and Thoms (1987) has also documented the changing morphology of a channel confluence in response to upstream river regulation. However, the change in bedform at the Widdale/Ure confluence had a significance that went beyond the temporary storage of so much material *in* the confluence precinct itself. There was the question of its effect on subsequent flows. The new bar was armoured with clasts that were, on average, twice the size of the armour in both contributing streams. This extremely well-armoured obstacle to flow was a persistent feature of the confluence for up to two years after the event that laid it. It was undoubtedly contributary, through deflection of the flow, to the subsequent bank erosion that occurred opposite the Ure entry.

Turning to the ephemeral drainage system under consideration, it is possible to see an exaggeration of the depositional consequences of out-of-phase floods. Because all rain falls from discrete convective cells in tropical semi-deserts, the probability of flow in two confluent streams of any size is low. The probability of coincident or near-coincident flood waves is even lower. This is reflected in the channel fill of the Il Warata/Il Naitiwa confluence (Figure 8.7).

The distinctly different mineralogy of the sediment carried by each stream allowed a four-fold classification of each bed that made up the channel fill: i.e. Warata, Naitiwa, mixed but Warata dominated, and mixed but Naitiwa dominated. At the confluence itself (Trench 1, Figure 8.7), pure beds of Warata sediment dominated, not only towards the right bank—the side of the Warata

Floods and Flood Sediments at River Confluences 145

Figure 8.6 (a) Bedforms at the River Ure/Widdale Beck confluence before 6–7 June 1982 flood; (b) bedforms after the 6–7 June flood in which the discharge ratio between the Ure and Widdale reached a maximum value of 23.

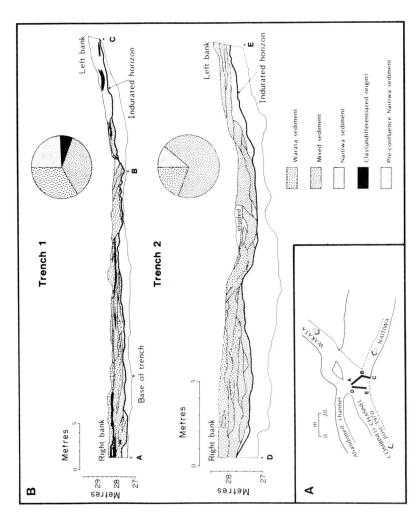

Figure 8.7 Trench sections of the channel fill at the Il Warata/Il Naitiwa confluence. The provenance of material making up individual beds was decided on the basis of distinctive clast mineralogy. The pie diagrams indicate the proportion of the exposure lying above the indurated horizon that marks the base of the confluence deposit identified as originating from different catchment sources.

entry—but also well over towards the left bank. They made up 33% of the total exposure. Even so there was one pure bed of Naitiwa sediment towards the right bank indicating that there had been a flow from this stream that had not been matched by flow from the Warata. Just downstream (Trench 2, Figure 8.7), the channel fill had become dominated by beds of variably mixed sediment. However, there were still pure beds of either Naitiwa or Warata sediment, though the proportion of the fill that they occupied had fallen from 58% to 30%.

The pure beds of Naitiwa sediment represent a discharge ratio of zero, while those of the Warata represent a ratio value of infinity. Neither of these extremes are encountered, of course, in perennial drainage systems such as the Ure/Widdale.

CONCLUSION

Confluences are important elements of the drainage network and are points at which flood flows are extremely complex. Nevertheless, several distinct patterns emerge from the discharge records collected at the confluence of two perennial streams. Each pattern is a function of either catchment conditions antecedent to a runoff generating storm, or the uneven distribution of rainfall over contiguous drainage basins. In the absence of large differences in lithologies in the contributing areas of each stream and substantial and unequal changes in land use, the intrinsic catchment variables are either seasonal, relating to soil moisture deficits, or topographical. Because of this, they are either 'fixed', or if climatic data are available, they vary predictably. On the other hand, the extrinsic factor—rainfall distribution—is not predictable. In a temperate setting like the British Isles, the frequency of discrete convective rainfall is low. In contrast, *all* rain falls from convective cells in tropical settings such as East Africa. Here, the probability that flow will occur in two confluencing rivers simultaneously is low, while the probability of flood waves entering a confluence from each stream in phase is even lower.

The non-coincidence of flood flows will be reflected in both the bed morphology and depositional facies. Extensive and rapid changes in bed morphology can be expected within the confluence precinct as a consequence of changing sediment transport patterns as the discharge ratio varies through one flood event. This will also be reflected both in the complexity of the depositional forms, such as avalanche faces which migrate into or recede from the confluence, and in the composition of sediments which may be represented at these sites. This holds important implications for the palaeohydraulic interpretation of channel confluences in ancient sediments.

ACKNOWLEDGEMENT

We are indebted to the Natural Environment Research Council for funds, both in the form of a research grant and a research training award.

REFERENCES

Alam, M. M., Crook, K. A. W., and Taylor, G. (1985). Fluvial herring-bone cross-stratification in a modern tributary mouth bar, Coonamble, New South Wales, Australia, *Sedimentology*, **32**, 235–44.

Ashmore, P. E. (1979). *Laboratory Modelling of Braided Streams*, Unpubl. M.Sc Thesis, Department of Geography University of Alberta, 195 pp.

Ashmore, P. E. (1982). Laboratory modelling of gravel braided stream morphology, *Earth Surface Processes*, **7**, 201–25.

Ashmore, P. E., and Parker, G. (1983). Confluence scour in coarse braided streams, *Water Resources Research*, **19**, 392–402.

Behlke, C. E. (1964). *An Investigation of Supercritical Flow Channel Junctions*, Engineering Experimental Station, Oregon State University Contract Report, No. CPR 11-7996, Bureau of Public Roads, U.S. Dep. Commerce, 59 pp.

Behlke, C. E., and Pritchett, H. D. (1966). The design of supercritical flow channel junctions, *Highway Research Record, Highway Research Board. Division Engineering, National Research Council Report*, **123**, 17–35.

Best, J. L. (1985). *Flow Dynamics and Sediment Transport at River Channel Confluences*, unpublished Ph.D Thesis, Birkbeck College, University of London, 393 pp.

Best, J. L. (1986). The morphology of river channel confluences, *Progress in Physical Geography*, **10**, 157–74.

Best, J. L. (1987). Flow dynamics at river channel confluences: implications for sediment transport and bed morphology. In: Recent Developments in Fluvial Sedimentology, *Society Economic Paleontologists and Mineralogists Publication*, **39**, 27–35.

Best, J. L. (1988). Sediment transport and bed morphology at river channel confluences, *Sedimentology*, **35**, 481–498.

Best, J. L., and Brayshaw, A. C. (1985). Flow separation—a physical process for the concentration of heavy minerals within alluvial channels, *Journal Geological Society*, London, **142**, 747–55.

Best, J. L., and Reid, I. (1984). Separation zone at open channel junctions, *Proceedings American Society Civil Engineers, Journal Hydraulic Engineering*, **110**, 1588–94.

Best, J. L., and Reid, I. (1987). Separation zone at open channel junctions: Closure, *Proceedings American Society Civil Engineers, Journal Hydraulic Engineering*, **113**, 545–8.

Black, P. E. (1970). Runoff from watershed models. *Water Resources Research*, **6**, 465–77.

Bowers, C. E. (1970). *Hydraulic Model Studies for Whiting Field Naval Air Station, Part 5: Studies of Open-Channel Junctions*, U.S. Department of Agriculture and St Anthony Falls Hydraulic Laboratory Project Report 24, 61 pp.

Frostick, L. E., and Reid, I. (1977). The origin of horizontal laminae in ephemeral stream channel fill, *Sedimentology*, **24**, 1–9.

Frostick, L. E., and Reid, I. (1979). Drainage net control of sedimentary parameters in sand-bed ephemeral streams, in A. F. Pitty, (ed.), *Geographical Approaches to Fluvial Processes*. GeoAbstracts, Norwich, 173–201.

Frostick, L. E., Reid, I., and Layman, J. T. (1983). Changing size distribution of suspended sediment in arid-zone flash floods, in J. D. Collinson and J. Lewin, (eds), *Modern and Ancient Fluvial Systems. Special Publication International Association Sedimentologists,* **6**, 97–106.

Gildea, A. P., and Wong, R. F. (1967). Flood control channel hydraulics, *Proceedings 12th International Association Hydraulic Research Congress, Fort Collins,* **1** (Paper A41), 330–7.

Itakura, T. (1972). Study of the mechanism of flow at confluences (in Japanese). *Proceedings 16th Hydraulics Conference Japanese Society Civil Engineers,* 7–12.

Jaeggi, M. N. R. (1986). Non distorted models for research in river hydrology, *Proceedings of the International Association for Hydraulic Research, Symposium on Scale Effects,* Toronto, 70–85.

Kennedy, B. A. (1984). On Playfair's law of accordant junctions, *Earth Surface Processes and Landforms,* **9**, 153–73.

Knighton, A. D. (1980). Longitudinal changes in size and sorting of stream-bed material in four English rivers, *Bulletin Geological Society America,* **91**, 55–62.

Knighton, A. D. (1982). Longitudinal changes in the size and shape of stream bed material: evidence of variable transport conditions, *Catena,* **9**, 25–34.

Kochel, R. C., and Baker, V. R. (1982). Paleoflood hydrology, *Science,* **215**, 353–61.

Komura, S. (1973). River-bed variations at confluences, *Proceedings Symposium River Mechanics, Bangkok,* 9–12 January, Paper A66, 773–84.

Krigström, A. (1962). Geomorphological studies of sandur plains and their braided rivers in Iceland, *Geografiska Annaler,* **44**, 328–46.

Lodina, R. V., and Chalov, R. S. (1971). Effect of tributaries on the composition of river sediments and of deformations of the main river channel, *Soviet Hydrology,* **4**, 370–4.

McGuirk, J. J., and Rodi, W. (1978). A depth averaged mathematical model for the near field of side discharges into open channel flow, *Journal of Fluid Mechanics,* **86**, 761–81.

Miller, J. P. (1958). *High Mountain Streams: Effects of Geology on Channel Characteristics and Bed Material,* State Bureau of Mines and Mineral Resources New Mexico Institute of Mining and Technology, Socorro, New Mexico, Memoir No. 4.

Mosley, M. P. (1975). *An Experimental Study of Channel Confluences,* Unpublished Ph.D Thesis, Colorado State University, 216 pp.

Mosley, M. P. (1976). An experimental study of channel confluences, *Journal of Geology,* **84**, 535–62.

Mosley, M. P. (1982). Scour depths in branch channel confluences, Ohau river, Otago, New Zealand, *Proceedings New Zealand Institute Professional Engineers,* **9**, 17–24.

Mosley, M. P., and Schumn, S. A. (1977). Stream junctions—a probable location for bedrock placers, *Economic Geology,* **72**, 691–4.

Muskatirovic, J., and Miloradov, M. (1980). Some experiences in improving river confluences, *Proceedings Symposium on River Engineering and its Interaction with Hydrological & Hydraulic Research, Belgrade, Yugoslavia* Paper D1, 11 pp.

Petts, G. E., and Thoms, M. C. (1986). Channel Aggradation below Chew Valley Lake, Somerset, U.K., *Catena,* **13**, 305–20.

Petts, G. E., and Thoms, M. C. (1987). Morphology and Sedimentology of a Tributary Confluence Bar in a Regulated River: North Tyne, U.K., *Earth Surface Processes and Landforms,* **12**, 433–40.

Prohaska, S. J. (1978). Hydrologic flood routing in confluence reaches, *Proceedings International Conference Water Resources Engineering, Bangkok, Thailand,* 689–702.

Reid, I., and Frostick, L. E. (1987). Flow dynamics and suspended sediment properties in arid zone flash floods, *Hydrological Processes,* **1**, 239–53.

Richards, K. S. (1980). A note on the changes in channel geometry at tributary junctions,

Water Resources Research, **16**, 241–4.
Rose, J. (1980). Landform development around Kisdon, upper Swaledale, Yorkshire, *Proceedings Yorkshire Geological Society,* **43**, 201–19.
Roy, A. G., and Roy, R. (1988). Changes in channel size at river confluences with coarse bed material, *Earth Surface Processes & Landforms,* **13**, 77–84.
Roy, A. G., and Woldenberg, M. J. (1986). A model for changes in channel form at a river confluence, *J. Geology,* **94**, 402–11.
Sharon, D. (1974). The spatial patterns of convective rainfall in Sukumaland, Tanzania— a statistical analysis, *Arch. Met. Geoph. Biokl. Ser. B,* **22**, 201–18.
Taylor, E. H. Jr. (1944). Flow characteristics at open channel junctions, *Transactions American Society Civil Engineers,* **109**, Paper 223, 893–912.
US Corps of Engineers (1970). *Hydraulic Design of Flood Control Channels,* Department of the Army, Corps of Engineers, Engineering and Design Manual, EM 1110-2-1601, 66 pp.

9 Flood Effectiveness in River Basins: Progress in Britain in a Decade of Drought

MALCOLM D. NEWSON

Professor of Physical Geography, University of Newcastle-upon-Tyne, Newcastle-upon-Tyne

INTRODUCTION

During the last twenty years geomorphologists have discussed at length the role of floods in developing the landscape under two main headings: 'work' and 'effectiveness' (Wolman and Gerson, 1978; Newson, 1980). Although 'work' was a natural obsession of the 1960s and early 1970s with the widespread monitoring of river flows and sediment loads, it was quickly realised that during very big events most measuring systems fail to cope. Thus, since retrospective flood reconstructions are required, these very big events have become dominant in the geomorphological literature by dint of the number of 'effectiveness' papers appearing in the 1980s. These papers record flood levels, erosion features, sediment size, deposition patterns and so on for their own sake but also because they are steps on the way to hydrological and hydraulic interpretations of the magnitude and frequency of the event.

It must be admitted that many geomorphologists have a basic fascination with flood effectiveness; they flock to the scene of any new cataclysm as a reward for merely observing years of 'normal' conditions. There is also a more cerebral interest: we dwell with Hutton, Lyell and Buckland when we contemplate the effectiveness of rare floods. In most of the small, remote, upland valleys, where such floods make their biggest impact, one can escape the present-day need for obeisance to funding authorities and faculty boards and refute Chorley's (1978) claim that we have exchanged teleology for conventionalism in geomorphology. Lower down in the same basin, on the floodplains, quite a different record is left too; can the two forms of research framework and the two locations be logically united to give basin-scale flood studies a more integrated and influential role? This is a question returned to after using a rather biased, British Isles selection of floods to read the tea leaves.

FLOOD STUDIES IN BRITAIN—AN ABBREVIATED BACKGROUND

It is a common feature of human response to hazard that resources are only applied to hazard studies immediately after a devastating or damaging event or, more likely, a series of them. Arnell *et al.* (1984) have pointed this out in connection with flood insurance in the UK, but it is a feature of all aspects of flood studies.

Thus, it was the dam failures of the late nineteenth century and the spectacular and fatal disaster at the Dolgarrog dams in 1925 (Fearnsides and Wilcockson, 1928) which led to the foundations of engineering flood hydrology in Britain in the form of the Institution of Civil Engineers report of 1933. The purpose of this report was to provide technical back-up to the Reservoir (Safety Provisions) Act of 1930. By coincidence it also heralded what were at the time the three driest years on record in Britain (1933, 1934, 1935) and therefore reservoir yield studies did not linger long in the shadow of reservoir flood hydrology.

Curiously a very similar pattern can be detected in the next major addition to the compendium of flood predictions. The *Flood Studies Report* (Natural Environmental Research Council, 1975) resulted from a study at the Institute of Hydrology, commenced in 1970 as a direct result of extensive flooding in Southern Britain in 1968. The envelope curve, or experience curve, of the 1930s had become outdated by two scientific developments, although it was still in use to check out the relative magnitudes of certain devastating natural floods such as that at Lynmouth in 1952 (see Dobbie and Wolfe, 1953). The two scientific developments were closely linked: data and statistics. Hydrological data collection in Britain had begun in earnest as the result of the Water Resources Act, 1963, though much of the instrumentation was designed for low flows, not rare floods. Nevertheless, it appeared very primitive to be using experience curves to deal with as wide a range of design problems as spillways and land drainage when each class of problem clearly involved its own levels of risk and hence probability. The reconciliation of increasing data and maturing analysis might, however, have waited decades without the flooding of 1968.

The year of publication of the *Flood Studies Report* was very dry! The following year, 1976, was even drier (Doornkamp *et al.*, 1980) and by the time the 1984 drought affected the north of the country (Marsh and Lees, 1985), low-flow studies had once again emerged as the preoccupation of many hydrologists (Institute of Hydrology, 1980). Morris and Marsh (1985) had indicated that the reduced summer rainfall *totals* of the decade 1975–84 represent real climatic change.

Flood Effectiveness in River Basins

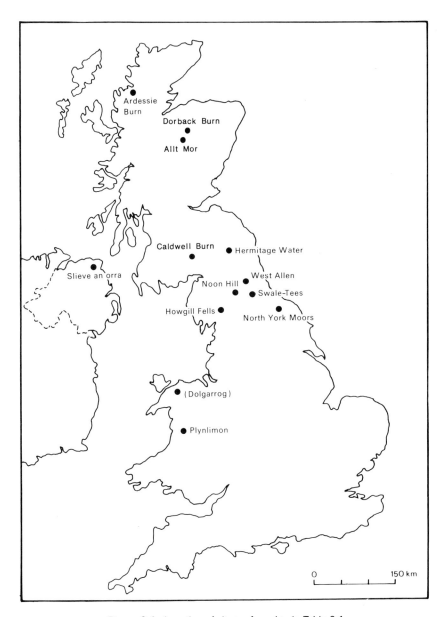

Figure 9.1 Location of sites referred to in Table 9.1.

Table 9.1 Notable examples of headwater flood effectiveness during a decade of drought.

Authors	Location	Date	Area (km²)	Rain	Q_p (cumecs)	Frequency (yrs) P-Rainfall Q-Flow	Effectiveness
Beven et al., 1978	N. York Moors	9–11.9.76	n.a.	86mm/80min	n.a.	—	Landslip/debris flow
Newson, 1980	mid-Wales	15.8.87	8.7	98mm/360mins	34	$P>100$ $Q<1000$	Channel deposition
Werritty, 1980	Allt Mor, Invernessshire	4.8.78	16.4	33.5mm/60mins	60	P 20–30	Channel change/ boulder transport
Werritty, 1984	Dorback Burn, Invernessshire	6.6.80	18.6	n.a.	80	—	Bank erosion/ Chaotic deposition
Tomlinson and Gardiner, 1982	Slieve an Orra Co. Antrim	1.8.80	n.a.	97mm/45min	n.a.	$Q>100$	Bog slides
Acreman, 1984	Ardessie Burn, Wester Ross	20.9.81	13.3	140mm/day	60	P 120	Debris flow/ channel avulsion
Brown (personal communication)	West Allen, Northumberland	17.7.83	88	25mm/90mins	67	Q 3.25	Scour/channel change
Carling, 1986a, b; 1987	Noon Hill, Northumberland	17.7.83	13	105mm/150mins	32	P 400–2500 Q 8–20	Peat slides Boulder-berms, -jams
Harvey, 1986 Wells and Harvey, 1987	Howgill Fells, Cumbria	6.6.82	1–8	70mm/150mins	8–30	P 100–500	Slides, flows, fans, Channel metamorphosis
Werritty, unpublished	Caldwell Burn, Dumfriesshire	13.6.79	5.6	90mm/60mins	190	$P/Q>1000$	Small slope failures, Scour
	Hermitage Water, Roxburghshire	25.7.83	36.1	64mm/75mins	170	$Q>500$	Many slides, flows, fill, metamorphosis
Macklin and Newson (in preparation)	Swale/Tees, Yorkshire/ Durham	26.8.86	2–86	118/day	2–200	$P<100$	Scour, chaotic deposition, run-out

FLOOD EVENTS DURING THE GREAT BRITISH DROUGHT

Climatic variability offers the analyst very conflicting signals according to the filter imposed on the total data-base. The great British drought is a clear signal at an annual or seasonal level. Rainfall amounts are, however, increasing over shorter timescales (Perry and Howells, 1982; 1986). Thus we have not been *free* of flooding in Britain during a decade of drought, though we have been free of *disastrous widespread* flooding. Possibly the most important flood during the period was that which flooded South Wales in December 1979. The main damage was caused in Cardiff, where a comprehensive flood relief scheme was proposed in the same month (HRS, 1979) and where eventually the flood warning system was officially criticised (Welsh Water Authority, 1981). This event, and others in Wales and in the lower Severn in subsequent winters (notably 1984) have thrown up again the inadequate attention we give to snow-melt events in Britain, though not for want of scientific warnings from northern England (Archer, 1981; 1983). Within Britain there are spatial patterns in contemporary climatic variability. As Figure 9.1 shows, northern Britain has played a prominent role in 'rare' floods since 1975 (see Table 9.1). All of these floods occurred in summer or early autumn and, in at least two cases, preceding drought is ventured on as an exacerbating factor. Another common factor in these accounts is their concentration on geomorphological effectiveness; there are clear dangers of bias, therefore, in my selection—how many remote upland valleys passed similarly 'extreme' events without bearing the scars (as in the case of Werritty's Dumfriesshire example). Nevertheless, effectiveness and recovery have been recurring research themes in this sample of floods, possibly encouraged by the lead provided by Anderson and Calver (1977; 1980). Clearly, therefore, the rash of papers from the north does not represent regional deprivation from flood protection by official agencies. It represents the greater likelihood of heavy rainfall affecting remote, rural, drift-covered, steep valleys in that part of Britain. There are possibly, therefore, two further compelling reasons for the northern domination of flood accounts: frequency of catchments of approximately 10 km^2 area (Baker and Costa, 1987) and a growing number of flood observers. Southern factors also operate: flood protection schemes in the lowlands of the south have been effective during the period; the experiences of Cardiff in 1979 stands out as an example of lack of such protection. If regional deprivation is to be claimed in terms of flood *reporting* it is Ireland which has justifiable claims. Whilst Tomlinson and Gardiner's (1982) report is entered in Table 9.1 because of its rainfall, discharge and frequency data, the many other bog bursts, bog flows and bog slides of the Emerald Isle are seldom reported in detail (but see Tomlinson, 1981; Alexander *et al.*, 1986; Coxon *et al.*, Chapter 12 in this volume). In fact the periodic excavation of large areas of organic terrain is the most complete demonstration of geomorphological thresholds. In an unpublished study of a decade of recovery from the 1973 flood on Plynlimon (see

Figure 9.2 'Burst' feature formed during the Plynlimon floods of 5-6 August 1973.

Flood Effectiveness in River Basins 157

Figure 9.3 The same 'burst' photographed ten years after its formation, revegetated and with peat regrowing.

Newson; 1980) the author recorded sites on which peat was actively reforming (see Figures 9.2 and 9.3), despite complete evacuation from a flush area during the event.

CONCEPTUAL PROGRESS IN FLOOD EFFECTIVENESS STUDIES OVER THE PERIOD

Up in the hills

The 1973 and 1977 floods on Plynlimon suggested that only rarely did the whole 'jerky conveyor belt' of sediment transport systems in river basins become active in one event. Long, low-intensity rainstorms can wet *slopes* to the point of failure, but cannot produce flood peaks to wreck channels. Storms of higher intensity produced rapid slope runoff and large channel flows, wrecking low-order channels, but are maintained for too short a period to transport large quantities of material out to piedmont and lowland reaches. The Lynmouth flood of 1952 was remarkable for being a 'slope flood' *and* a 'channel flood'. In Table 9.1 it is clear that whilst some of the entries support this division others refute it (or represent events of a rarity similar to Lynmouth). The Noon Hill (Carling 1986a,b; 1987) and Howgill (Harvey 1986; Wells and Harvey, 1987) events clearly combined slope and channel effectiveness, whilst the Hurricane Charley floods in the Northern Pennines illustrate mainly channel effectiveness but with a very impressive downstream 'run out' of finer sediments across floodplains.

Are these floods rarer than those which merely affect slopes or channels? The evidence from Table 9.1 is to the contrary; in fact Table 9.1 suggests that all future comment on flood return periods should be curtailed until the figures mean something! The *Flood Studies Report* (Natural Environment Research Council, 1975) must not be used as a platform for spurious accuracy! Possibly the most accurate frequency quoted in the table is the 3.25-year flood which achieved so much channel effectiveness on the West Allen, because of its rapidity of rise (Brown, personal communication). Werritty's (personal communication) figure for the Caldwell Burn flood is in direct contrast: a return period of over 1000 years, achieving very little effectiveness, for reasons which the same author delves into in a 1984 paper. Clearly, an *effectiveness-frequency model* will be longer in development than was a work–frequency model, and it might be as well to concentrate on the insights these floods give in into *sediment transport processes* during the 'big event' (I cannot persist with 'X-year floods'; Baker and Costa, 1987, use 'rare, great floods' for those effective across slopes and channels or down impressively long reaches of the channel network).

Following disasters in the United States, both natural (e.g. Mount St Helens volcano) and man-made (e.g. Lawn Lake and other dam disasters) the decade has seen a wealth of material gathered on flow processes, including important evidence of non-Newtonian flows and sediment transport.

Central to the study of these very effective floods has been the interpretation of the power and process by which the impressive sediment transport feats of these events are achieved, (Baker and Costa, 1987). As a crucial avenue of pure research we see this set up as early as 1928 in the discussion of the Dolgarrog paper by Fearnsides and Wilcockson (E. H. L. Schwarz is the discussant).

> I am trying to find some means by which large rocks, such as we have seen on the screen, can be carried. Was it pure water that moved them, or was it not rather a sort of mud-rush; that the water was full of mud and clay, and that the specific gravity of the fluid was not 1, but something approaching 2 and $2\frac{1}{2}$, and under those circumstances it could float big stones?

The 'mud-rush' hypothesis has steadily become consolidated and the most recent collation of evidence for the importance of debris flows has come from Costa (1984); of even greater impact has been the video collation of debris flows in action (Costa and Williams, 1984). The vast admixture of fine sediments introduced to rivers following the Mount St Helens eruption of 1980 has forced a reclassification of sediment-laden flows (Table 9.2) in terms of triggering mechanisms, sediment concentrations and flow processes (see Bradley, 1986, for a field and laboratory study). What emerges is a complicated picture of density, viscosity and flow resistance effects, all tending to increase sediment transport for a given flow. The most important transition is clearly that at which flow processes become non-Newtonian and therefore laminar, uniform and with very low particle settling velocities. Such flows are often the best process to invoke for certain flood depositional forms such as boulder berms (Table 9.3 lists certain characteristics of these features). However, one must exercise caution in two respects. One is that there clearly has to be a source of fines within the flood source area. Blair (1987) has pointed out that in the case of the Lawn Lake dam-break flood there was no such source; boulder levees and trains resulted, respectively, from an initial 'non-cohesive sediment gravity flow' (NCSGF), controlled by an upstream gorge section, and subsequent sheet flooding characterised by large amounts of tree debris. The sedimentary evidence of the Dolgarrog disaster concurs with Blair's interpretation: boulder berms alternate with sheet flood deposits according to valley morphology and sediment sources (Figures 9.4 and 9.5). Debris flow and torrents are almost certainly restricted to channels 'coupled up' with neighbouring slopes and gulleys during the flood event. NCSGFs are possibily the more common phenomena, resulting from valley constrictions, especially where there are river cliffs of glacial drift.

Wells and Harvey (1987) have used the geographical and sedimentological range of flood deposits in the Howgill Fells to add a further note of caution to

Table 9.2 Classification of Flow with High Sediment Concentrations (Bradley, 1986).

Source	Concentration percent by weight (100% by WT = 1000000 ppm)									
	23	40	52	63	72	80	87	93	97	100
	Concentration percent by volume (S. G. = 2.65)									
	10	20	30	40	50	60	70	80	90	100
Beverage and Culbertson (1964)	High	Extreme					Mud flow			
Costa (1984)	Water flood		Hyperconcentrated					Debris Flow		
O'Brien and Julien (1985) using National Research Council (1982)	Water flood		Mud Flood				Mud Flood	Landslide		
Takahashi (1981)	Fluid flow				Debris or Grain Flow			Fall, Landslide, Creep, Sturzstrom, Pyroclastic flow		
Chinese Investigators (Fan and Dou, 1980)	Sediment laden				----- Debris or Mud Flow -----					
					----- Hyperconcentrated Flow -----					
Pierson & Costa (1984)	Streamflow Normal: Hyperconcentrated				Slurry flow (Debris torrent), Debris and mud flow, Solifluction			Granular flow Sturzstrom, Debris Avalanche, Earthflow, Soil creep		Fast / Slow

Table 9.3 Characteristics of boulder berms formed by debris torrents (Costa, 1984).

(1)	They form below rapidly expanding reaches.
(2)	They may not form on both sides of the channel.
(3)	They are short, but continuous; lengths range from 2 to more than 10 m.
(4)	They form below reaches with large sources of debris, such as deep scour or landslides.
(5)	They are mostly grain-supported with little matrix. The matrix is coarse and usually contains less than 4 to 5 percent silt and clay.
(6)	Boulders have steep imbrication angles, commonly greater than 60°.
(7)	Long axes are perpendicular to the flow direction.
(8)	Some of the coarsest rocks are on the top of the berms.
(9)	The tops of the berms may be above high-water marks on valley sides.

those arguing for the pervasive influence of debris flows. Within the range of alluvial fans in the Howgills there are signs that only smaller basins tended to exhibit debris flows (dilution becoming important downstream) and that, within some basins, flow and transport processes changed across the Newtonian/non-Newtonian threshold and back during the flood. The signals offered from such studies to geologists interpreting ancient flood deposits seem at first confusing. In fact they offer a discipline for such interpretations: always look upstream, always be prepared for mixed sedimentary evidence within small distances, always sample in detail and always attempt to reconstruct the surface upon which deposition occurred.

Furthermore, such studies have a more profound signal for earth scientists philosophising over the magnitude–frequency conundrum. As Baker and Costa (1987) have pointed out, 'Basins of approximately 10 km^2 seem to have the valley and hydraulic characteristics that optimise flood depth, energy slope and velocity to maximize flood power'. Costa (1987) adds, 'Maximum flood peaks originate from an optimal combination of basin morphology, physiography and storm intensity.' Werritty (personal communication) adds three important characteristics specifically localising maximum flood effectiveness to small basins—flow variability, sediment size, and thickness of drift/organic cover are all at their maximum. Such catchments also provide maximum likelihood of 'slope' and 'channel' floods combining and, therefore, the greatest likelihood of NCSGFs, debris flows and other forms of hyperconcentrated sediment transport mechanisms.

Down on the plains

In terms of overall morphological effect we must not allow 'rare, great floods' to make us dogmatic catastrophists beyond a basin area of, say, 10 km^2. In reaches where even the smallest floodplains are developed there is much less likelihood of dramatic effectiveness. Nanson's (1986) illustration of flood-plain stripping in

Figure 9.4 Boulder berm formed in the Dolgarrog flood (1925) below a constricted reach with boulder clay cliffs.

valleys confined by bedrock is an unusual extreme. Outwith confined reaches one might expect the impact of such floods on most floodplains to be one of accretion, with possible subtle channel adjustment, whilst an increased loading of in-channel sediment is distributed by scour and fill (Pitlick, 1985). There may not even be flood sediments to route through such reaches: beyond the mountain

Figure 9.5 Sheet flood deposits at Dolgarrog in the wider reach immediately downstream of Figure 9.4.

margin, delivery ratios for flood sediments may be zero as demonstrated by Osterkamp and Costa (1986). Pitlick (1985) demonstrates that in reaches beyond major depositional sites, such as the Roaring River fan a meandering alluvial channel can pass the entire flood-imposed load of fine gravel and sand by scour-and-fill processes, without morphological change. A further interesting aspect of the Lawn Lake disaster was that below the stable reach Pitlick describes a further, minor, dam failure, creating another suite of devastated, depositional and stable reaches. Lake Estes curtails the sequence, but in large basins where this sequence develops fully (e.g. in British basins with polycylic long profiles) the pattern becomes similar to the alternation of depositional and transport reaches described for the Bella Coola River, British Columbia by Church (1983). The flood's effect on the remainder of the river system depends, therefore, on the initial coupling from source area to transition zone and, thereafter, may have an impact similar to that of glacial sediments in 'paraglacial' river basins (Church and Ryder, 1972).

There is an attractive hydrological corollary to the fact that reported flood events during the period have mainly been analysed in the context of geomorphological effectiveness. Hopefully such studies will eventually allow reflexive benefits for flood prediction. There remains, locked up in the geomorphological record, information of exactly the timescales relevant to extension of records beyond historical data collection (see Baker, Chapter 10 in the volume).

The surveys of Pennine valleys which followed the Hurricane Charley floods of August 1986 revealed that events of this magnitude had clearly occurred before (Figure 9.6). They can be dated by a combination of lichonometry and historical documentation to enhance flood frequency analyses around and beyond the crucial 100-year event. Whilst the 100-year flood is seldom a design problem in remote upland valleys, the extensive overbank deposition which Charley brought about lower down the river systems of the North (Figure 9.7) suggests that floodplain sediments in the neighbourhood of gauging stations, taken in conjunction with historic documentary records of flooding and channel change, may open up the true potential of a geomorphological contribution to flood frequency analyses.

Of course, there are problems of assuming stationarity of climate and stage-discharge curves with this approach. Geomorphologists cannot have it both ways: we cannot thrill to the subtleties of environmental change recorded by rivers (including those induced by Man) and yet claim to identify extensions of stable information beyond the instrumental record. We cannot, either, guarantee stationarity of extreme flood sedimentary signatures. Research described above has identified the wide range of process–response models which might apply to big events. Even the notion of separating 'slope' from 'channel' floods may mean that, as in ancient sediments, the gaps carry as much (lost) information as the layers either side. Even within the 'channel' group of floods (in terms of effectiveness) a wide range of sediment transport processes has been identified. Hurricane Charley was notable in northern England for the continuity of sedimentary effects down the river systems from massive sandstone boulders transported in the uplands to cobble fans at tributary junctions and thick sand layers across floodplains tens of kilometres downstream. How rare is such an event? Future consideration of the flood evidence of such suites of deposits will work on the hypothesis that such events may be relatively common in regions of similar lithology, morphology and climate to the Northern Dales. Lithology (both solid and drift) in principle may provide a full range of sediment sizes, through the system and record flood effectiveness. In the Palaeozoic shale uplands of Britain neither the bedrock nor the drift yield a wide particle-size range; gravels and silts dominate. In floods the gravels clog piedmont reaches whilst silts reach the ocean. Furthermore, the Northern Dales rivers transport a useful semi-natural tracer in the form of metaliferous mining waste. The applied value of investigating floodplain stratigraphy in the region has already led to excavations, designed to reveal the rate of pollutant release to crops and which, incidentally, also yielded a highly logical flood stratigraphy across low terraces of the Tyne. What of non-stationarity stemming from other sources? The Swansea team suggest that northern Britain is outside the region where increased daily rainfall

Flood Effectiveness in River Basins   165

Figure 9.6 Evidence of historical flood deposits (foreground), 2–4 m above the flood-torn channel of the Hurricane Charley flood in Swaledale, Yorkshire.

Figure 9.7 'Run out' of finer, sand deposits and accretion of floodplains was a downstream characteristic of the Charley floods. Metal-polluted sands in Swaledale, Yorkshire.

totals are increasing flood frequencies (Walsh *et al.*, 1982). Nevertheless, the region has experienced rapid land-use change along with most of upland Britain; so that a combination of climatic and land-use influences can measurably alter hydrograph response. The implication of such work is that we are living in a period of floodplain changes; incision may be rapid but accretion is a slow process over much of Britain because of our low sediment yields.

Can geomorphological evidence provide better flood frequency estimates than that of river flow measurement? Almost certainly not; but it should be noted that there are also changes under way in attitudes to hydrological data collection on floods. A major effect of the economic problems facing the UK during recent years has been a decline in hydrometry, both in terms of the gauging network (both rainfall and stream flow stations have been rationalised) and the time allotted to its servicing. Johnson (personal communication) records a 75% decline in the latter aspects in Northumbria. There are signs, too, in recent legislation and policy decisions of a reduced effort in archiving hydrometric data, though the quality and rapidity of publication of data are both improving (Rodda and Monkhouse, 1984).

Nevertheless, it is clear that a prevailing policy stance is that only 'the best' river gauging stations should survive and have their data archived. Food warning systems are increasingly based on improved telecommunications and mathematical modelling. Flood design can be aided by the *Flood Studies Report* rather than by local gauging. In terms of the study of very large events it must be admitted that few gauging stations will cope and that indirect estimates of peak flow have always been necessary; nevertheless, the loss of 'ground truth' gauged data is serious.

Reduced expenditure on hydrometry is often justified in operational and financial terms. The biggest potential problem is that of research use of data from the hydrometric network, especially considering contemporary floods research problems which centre upon the need for duration and spread of data.

CONCLUSION

This review demonstrates the value of a determined and coordinated campaign by all relevant disciplines, but particularly by hydrologists and geomorphologists, to record the impacts of contemporary floods and to continuously review those of historical floods (and even prehistoric floods). Attempts at adding the vital co-ordination, such as the Royal Scottish Geographical Society initiative of the 1950s and 1960s and the BGRG Floods Patrol (1970s) have tended to fail. Possibly the best framework will remain that of special meetings of the relevant learned societies and the publication of proven standard procedures by experienced flood recorders (e.g. Williams and Costa, 1987). Nevertheless, published studies of flood effectiveness seldom encompass whole river systems.

The geomorphologist falls easily into the engineer's trap of taking a reach approach. For the study of flood effectiveness to be effective itself, much more comprehensive approaches are needed, using large multi-disciplinary teams and maintaining the study for as long as data remain indicative of this or that pattern of response. Without such basin-wide efforts, our research advances are likely to be as 'uncoupled' as the sedimentary transport system of smaller floods. The resources required for treating flood effectiveness at a basin scale are huge; the temptation is to record the 'rare great floods' and leave the rest. In the study of the recovery of the Roaring River/Fall River system, after the Lawn Lake Dam disaster in Colorado, no fewer than four funding agencies and three teams were involved, even beside the individual efforts of Jarrett and Costa (1984) and Blair (1987). Inevitably resources will only be available for the comprehensive study of flood geomorphology at times of crisis or where ancilliary problems such as floodplain pollution arise.

In view of the meagre resource base normally prevailing it seems that society gets extraordinarily good value from geomorphologists drawn irresistably to this most fascinating of hazards.

REFERENCES

Acreman, M. C. (1984). The significance of the flood of September 1981 on the Ardessie Burn, Wester Ross, *Scott. Geog. Mag.,* **100**, 150–60.

Alexander, K. W., Coxon, P., and Thorn, R. H. (1986). A bog flow at Straduff townland, County Sligo, *Proc. Royal Irish Acad.,* **86**(BA) 107–19.

Anderson, M. G., and Calver, A. (1977). On the persistence of landscape features formed by a large flood, *Trans. Inst. Brit. Geogrs.,* **2**, 243–54.

Anderson, M. G., and Calver, A. (1980). Channel plan changes following large floods, in R. A. Cullingford, D. A. Davidson and J. Lewin (eds.), *Time Scales in Geomorphology*, Wiley, Chichester, 43–52.

Archer, D. R. (1981). Severe snowmelt runoff in Northeast England and its implications, *Proc. Instn. Civ. Engrs.,* Pt. 2, **71**, 1047–60.

Archer, D. R. (1983). Computer modelling of snowmelt flood runoff in Northeast England, *Proc. Instn. Civ. Engrs.,* Pt. 2, **75**, 155–73.

Arnell, N. W., Clark, M. J., and Gurnell, A. M. (1984). Flood insurance and extreme events: the role of crisis in prompting changes in British institutional response to flood hazard, *Applied Geography,* **4**, 166–81.

Baker, V. R., and Costa, J. E. (1987). Flood power, in L. Mayer and D. Nash, (eds), *Catastrophic Flooding*, The Binghamton Symposia in Geomorpholgy: International Series, no. 18, 1–24.

Beven, K., Lawson, A., and McDonald, A. (1978). A landslip/debris flow in Bilsdale, North York Moors, September 1976. *Earth Surface Processes,* **3**, 407–19.

Blair, J. C. (1987). Sedimentary processes, vertical stratification sequences, and geomorphology of the Roaring River alluvial fan, Rocky Mountain National Park, Colorado, *J. Sed. Petrol.,* **57**(1), 1–18.

Bradley, J. B. (1986). *Hydraulics and Bed Material Transport at High Fine Suspended Sediment Concentrations.* Unpublished Ph.D. Thesis, Dept. of Civil Engineering, Colorado State University, Fort Collins.

Carling, P. A. (1986a). Peat slides in Teesdale and Weardale, Northern Pennines, July 1983: description and failure mechanisms, *Earth Surface Processes and Landforms,* **11,** 193–206.

Carling, P. A. (1986b). The Noon Hill flash floods; July 17 1983. Hydrological and geomorphological aspects of a major formative event in an upland catchment, *Trans. Inst. Br. Geogr.,* **NS11,** 105–18.

Carling, P. A. (1987). A terminal debris-flow lobe in the northern Pennines, United Kingdom, *Trans. R. Soc. Edinburgh,* **78,** 169–76.

Chorley, R. J. (1978). Bases for theory in geomorphology, in C. Embleton, D. Brunsden and D. K. C. Jones, (eds) *Geomorphology. Present Problems and Future Prospects,* Oxford Univ. Press, pp. 1–13.

Church, M. (1983). Pattern of instability in a wandering gravel bed channel, *Spec. Pubs. Int. Ass. Sediment,* **6,** 169–80.

Church, M., and Ryder, J. M. (1972). Paraglacial sedimentation: a consideration of fluvial processes conditioned by glaciation, *Geol. Soc. Amer. Bull.,* **83,** 3059–72.

Costa, J. E. (1984). Physical geomorphology of debris flows, in J. E. Costa and P. J. Fleisher, (eds.), *Development and Applications of Geomorphology,* Springer-Verlag, Berlin, pp. 268–317.

Costa, J. E. (1985). Floods from dam failures, *United States Geological Survey, Open File Rept.,* 85–560.

Costa, J. E. (1987). Hydraulics and basin morphometry of the largest flash floods in the conterminous United States, *J. Hydrol.,* **93**(3/4), 313–38.

Costa, J. E., and Williams, G. P. (1984). Debris flow dynamics (video recording), *United States Geol. Survey Open File Rept.,* 84–606, Denver, Colorado.

Dobbie, C. H., and Wolfe, P. O. (1953). The Lynmouth flood of August 1952, *Proc. Instn. Civ. Engrs.,* **2,** 522–88.

Doornkamp, J. C., Gregory, K. J., and Burn, A. S. (1980). *Atlas of Drought in Britain 1975–76,* Institute of British Geographers, London.

Fearnsides, W. G., and Wilcockson, W. H. (1928). A topographical study of the flood-swept course of the Porth Llwyd above Dolgarrog, *Geog. J.,* **72,** (s), 401–19.

Harvey, A. M. (1986). Geomorphic effects of a 100-year storm in the Howgill Fells, Northwest England. *Zeitschrift für Geomorphologie,* **30**(1), 71–91.

HRS (1979). Cardiff flood relief scheme, *HRS Rept. Ex 897,* Wallingford, Oxon.

Institute of Hydrology (1980). *Low Flow Studies,* Wallingford, UK.

Institution of Civil Engineers (1933). *Floods in Relation to Reservoir Practice,* London.

Jarrett, R. D. and Costa, J. E. (1984). Hydrology, geomorphology and dam-break modelling of the July 15, 1982, Lawn Lake dam and Cascade lake dam failures, Larimer County, Colorado. *United States Geological Survey, Open File Rept.,* 84–612.

Marsh, J., and Lees, M. (1985). *The 1984 drought,* Hydrological data UK. Institute of Hydrology/British Geological Survey, Wallingford, UK.

Morris, S. E., and Marsh, T. J. (1985). 'United Kingdom rainfall 1975–1984: evidence of climatic instability?' *J. Meteorology,* **10**(103), 324–32.

Nanson, G. C. (1986). Episodes of vertical accretion and catastrophic stripping: a model of disequilibrium flood-plain development, *Geol. Soc. Amer. Bull.,* **97,** 1467–75.

Natural Environment Research Council 1975. *Flood Studies Report,* London (5 Volumes).

Newson, M. D. (1980). The geomorphological effectiveness of floods—a contribution stimulated by two recent events in mid-Wales, *Earth Surface Processes,* **5,** 1–16.

Osterkamp, W. R., and Costa, J. E. (1986). Denudation rates in selected debris-flow basins, *Proc. 4th Federal Interagency Sedimentation Conference,* **1,** 4-91–4-99.

Perry, A. H., and Howells, K. (1982). Are large falls of rain in Wales becoming more frequent? *Weather,* **37**(8), 240–3.

Perry, A. H., and Howells, K. (1986). Changes in the magnitude-frequency of heavy daily rainfalls in the British Isles *Proc. 3rd Hellenic-British Climatol. Congress*, Athens, 1985, 49–54?

Pitlick, J. (1985). *The Effect of a Major Sediment Influx on Fall River, Colorado*. Unpublished M.Sc. Thesis, Colorado State University, Dept. of Earth Resources, Fort Collins, Colorado.

Rodda, J. C., and Monkhouse, R. A. (1984). The national archive of river flows and groundwater levels for the United Kingdom, *J. Instn. wat. Engrs. and Sci.*, **39**(4), 358–62.

Tomlinson, R. W. (1981). A preliminary note on the bog-burst at Carrowmaculla, Co. Fermanagh, November, 1979, *Irish Nat. J.*, **20**.

Tomlinson, R. W., and Gardiner, J. (1982). Seven bog slides in the Slieve-an-Orra Hills, County Antrim, *J. Earth Sci, Roy. Dublin Soc.*, **5**, 1–9.

Walsh, R. P. D., Hudson, R. N., and Howells, K. A. (1982). Changes in the magnitude-frequency of flooding and heavy rainfalls in the Swansea Valley since 1875, *Cambria*, **9**(2), 36–60.

Wells, S. G., and Harvey, A. M. (1987). Sedimentologic and geomorphic variations in storm-generated alluvial fans, Howgill Fells, Northwest England, *Geol. Soc. Amer. Bull.*, **98**, 182–98.

Welsh Water Authority (1981). *The Floods of 26–29 December 1979*, Brecon, Powys, UK.

Werritty, A. (1980). Allt Mor floods of 4th August 1978, *Brit. Geomorph. Res. Group Excursion Guide*.

Werritty, A. (1984). Stream response to flash floods in upland Scotland, in T. P. Burt and D. E. Walling, (eds), *Catchment Experiments in Fluvial Geomorphology*, Geo Books, Norwich, pp. 537–60.

Williams, G. P., and Costa, J. E. (1988). Geomorphic measurements after a flood, in V. R. Baker, R. C. Kochel and P. C. Patton, (eds) *Flood Geomorphology*, Wiley, N. Y.

Wolman, M. G., and Gerson, R. (1978). Relative scales of time and effectiveness in watershed geomorphology, *Earth Surface Processes*, **3**, 189–208.

10 Magnitude and Frequency of Palaeofloods

VICTOR R. BAKER
Department of Geosciences, University of Arizona, Tucson, Arizona

INTRODUCTION

Palaeoflood hydrology is the scientific study of past or ancient floods which occurred prior to the time of (a) direct measurement by modern hydrologic procedures, or (b) documentation by other human records. Type (a) flood records are also called instrumental records and are considered systematic for statistical analysis. Type (b) flood records are also called 'historical floods', and such records may or may not be systematic for various types of statistical analysis. The hydrological literature commonly treats historical floods and palaeofloods as equivalent. Such literature may also substitute the term 'systematic' for conventional instrumental records, thereby conveying the sometimes false implication that historical and palaeofloods are not systematic.

This paper will briefly review some recent research developments in the scientific study of palaeofloods. Because emphasis will be placed on the use of palaeoflood data in flood-frequency analysis, only the optimum palaeoflood hydrological procedure will be analysed in detail. There are three basic categories of palaeoflood hydrological data.

(1) Regime-based palaeoflow estimates (RBPE) involve empirically derived relationships that relate relatively high-probability flow events, such as the mean annual flood or bankfull discharge, to palaeochannel dimensions, sediment types, gradients, and other field evidence. These relationships apply to alluvial rivers, described by Leopold and Langbein (1962) as, '...themselves the authors of their own hydraulic geometries.'

(2) Palaeocompetence studies involve regression expressions and theoretical considerations that determine shear stress, velocity, or stream power (SS–V–SP) for maximum palaeofloods that transported very large sedimentary particles (Costa, 1983). Williams (1984) reviews the common equations used in both RBPE and SS–V–SP studies.

(3) Palaeostage-based flow data are generated from stable-boundary fluvial reaches characterized by the long-term preservation of slackwater deposits and

palaeostage indicators (SWD-PSI). In contrast to alluvial rivers, rivers displaying excellent SWD-PSI have been described by Baker and Pickup (1987) as, '...the chroniclers of their own cataclysms.' Because of their accuracy and completeness (Table 10.1), SWD–PSI studies are most amenable to magnitude–frequency analysis and will be the only type of palaeoflood hydrology to be considered further in this paper.

Table 10.1 Comparison of Palaeoflood Hydrological Techniques.

	RBPE	SS–V–SP	SWD–PSI
River type	Alluvial	Any	Bedrock
Magnitude	Moderate	Large	Large
Frequency	Moderate	Rarest	Rare and Rarest
Unit power	Low	Variable	High
Accuracy	Moderate	Low	High
Number of Floods	One	One or Few	Many

SWD-PSI

Slackwater deposits consist of sand and silt (sometimes gravel) that accumulate from suspension in extreme floods at local sites of reduced flow velocity. Stable-boundary channels are required, particularly for rivers with narrow–deep cross sections. Such channels commonly occur in resistant bedrock (Baker, 1984) and result in unusually great bed shear stress and power per unit area of bed (unit power) for great floods (Baker and Coast, 1987).

The Katherine Gorge in nothern Australia, described by Baker and Pickup (1987), illustrates an ideal situation for flood slackwater depostion. Stage changes of up to 15m occur during tropical floods, which emplace sandy deposits at the mouths of tributary canyons (Figure 10.1). Numerous slackwater deposits occur along any given reach of the gorge.

Because of their high loads of sand, arid-region bedrock streams may also provide excellent sites for SWD–PSI palaeoflood hydrologic analysis. In addition to tributary mouths, slackwater depositional sites may occur at abrupt channel expansions and constrictions, downstream of large flow obstructions, and in channel-margin caves and alcoves. Additional important palaeostage information can be provided by silt lines, flood scour lines, and flood-damaged vegetation. The latter are analysed extensively by Hupp (1988). O'Connor *et al.* (1986) and Partridge and Baker (1987) give examples of the uses of silt lines and scour lines respectively. Preservation times for such palaeostage indicators can vary from months to centuries.

An SWD–PSI palaeoflood hydrologic analysis proceeds in five major phases: (1) discovery in the field of key streams and individual sites that preserve appropriate SWD–PSI evidence; (2) stratigraphic and geochronologic analysis of individual flood-sedimentation units or palaeostage indicators in order to associate the palaeofloods with specific dates or time intervals; (3) hydraulic analysis of the chosen study reach; (4) relation of the SWD–PSI evidence to the channel hydraulics for the most accurate possible determination of palaeoflood discharges; (5) performance of a magnitude–frequency analysis of the data established in phases (2) and (4). This paper will review this procedure, emphasizing recent developments in phase (5).

STRATIGRAPHY

At ideal SWD–PSI sites, sequences of multiple sedimentation units record multiple palaeofloods, sometimes with intercalated non-flood units. Individual flood units and interbedded non-flood units or disconformities may be distinguished by a great variety of sedimentological features, including the following:

Figure 10.1 Sandy slackwater deposits and the intercalated tributary alluvium (coarse gravel) at the mouth of Cave Creek, a tributary to Aravaipa Canyon in southeastern Arizona. The uppermost sand was deposited by a flood of approximately $700 m^3 s^{-1}$ in October 1983. Organics at the top of the underlying gravel data at 1962 A.D. (Tx-5483) from the curve of Baker *et al.* (1985). The underlying flood deposit is associated with a formative discharge of at least $550 m^3 s^{-1}$. The organic layer underlying it, being pointed to by the student, has a radiocarbon date of 1958 A.D. (Tx-5484).

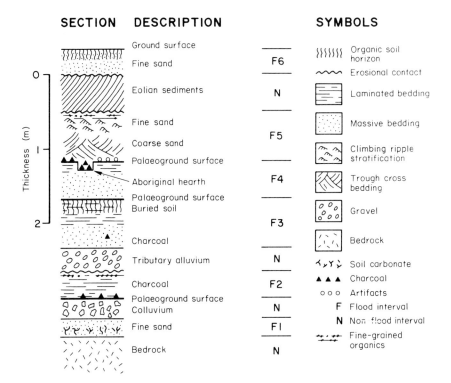

Figure 10.2 Idealized stratigraphic section of a slackwater sequence containing six flood deposits (numbered F1 to F6 at the centre of the figure) and several non-flood deposits (N). The symbols at right show the complexity of this hypothetical section. Note that several types of materials are present for geochronological analysis.

(1) silt–clay and organic drapes, (2) buried palaeosols, (3) organic layers, (4) intercalated tributary alluvium or slope colluvium, (5) abrupt vertical grain-size variations, (6) mud-cracks, (7) colour changes, and (8) induration properties. Figure 10.2 illustrates a complex SWD–PSI sequence displaying most of these properties. The interpretation and analysis of such sequences are reviewed by Baker (1987), Baker et al. (1983), Ely and Baker (1985), Kochel and Baker (1982, 1988), Kochel et al. (1982), and Partridge and Baker (1987). Two important stratigraphic concerns in SWD–PSI studies are completeness and interpretability of the record. Sites preserving slackwater records should show minimal erosion by tributary floods or slope processes. Another important concern is bioturbation, which destroys the stratigraphic breaks between flood-sedimentation units. Bioturbation problems are minimized in arid regions, in cave or rock shelter deposits, and in areas of rapid and thick sedimentation. As will be discussed below, problems of flood evidence missing at any one SWD–PSI site can be

minimized by stratigraphic correlation of flood layers to multiple sites, thereby establishing the complete flood record as a composite from many correlated SWD-PSI sites.

GEOCHRONOLOGY

Numerous recent advances have been made in the capability to associate precise ages with palaeoflood events. Great care must be exercised in all geochronological studies. The physical and chemical character of the system used for dating leads to one set of complexities. Another is posed by field relationships that make it difficult to place absolute ages in their proper context. A great variety of geochronological procedures may be used including: (1) diagnostic age-indicating materials such as mining debris or refuse; (2) archaeological materials; (3) dendrochronology; (4) thermoluminescence dating; (5) radiocarbon dating. The latter has been used most extensively, and some discussion of it will illustrate general principles.

The typical materials subjected to radiocarbon analysis in SWD-PSI studies are (1) organic materials in buried soils, (2) allochthonous charcoal or wood, (3) autochthonous charcoal, (4) flood-transported organic litter, and (5) buried trees. Problems of dating buried soils in flood sequences are discussed by Patton and Baker (1977). Allochthonous datable materials involve organics that are transported to the depositional site. Because such materials may be eroded from older alluvium, they can only provide minimum limiting ages on the flood sediments. A more useful stratigraphic association is with datable materials on the discontinuity surfaces that separate the flood layers. Where such materials were emplaced in the time interval between flood events they are termed autochthonous. They provide a precise minimum limiting age for the flood emplacing the immediate underlying slackwater deposit and a maximum limiting age for the flood emplacing the immediate overlying deposit. An exception to the problems of allochthonous materials occurs in cases where fine-grained organic detritus is preserved in the uppermost layers of an individual flood-slackwater unit. Seasonal gound litter transported by the flood may have a radiocarbon age within one year of the flood event that transported it (Baker *et al.*, 1985).

Recent developments in the radiocarbon dating of flood deposits include the 'post-bomb' dating of samples younger then 1950 A.D. Using appropriate curves of ^{14}C activity, it is possible to data samples to the precise year (Baker *et al.*, 1983; 1985). Even more important has been the development of accelerator mass spectrometry. The revolutionary advantage of this procedure is its very small required sample size (Taylor *et al.*, 1984). Individual seeds, blebs of charcoal and other small organic detritus can be dated with high accuracy.

PALAEOFLOOD MAGNITUDES

Recent work in SWD–PSI palaeoflood hydrology has emphasized accuracy in palaeodischarge determination. Computer flow models for step–backwater analysis are applied to the stable channel reaches containing SWD–PSI sequences. These models determine water-surface profiles for various hypothetical discharges that are routed through the reaches. By comparing the model-generated profiles to the SWD–PSI evidence (Figure 10.3) probable palaeodischarges are specified.

There are several strategies for reducing the errors inherent in this procedure. These are discussed for applications by Ely and Baker (1985), O'Connor *et al.* (1986), and Partridge and Baker (1987). One of the many useful procedures is the testing and calibration of the flow model with high-water mark evidence associated with relatively large flood flows that have been accurately measured at stream gauges (Ely and Baker, 1985).

STRUCTURING OF SWD–PSI DATA

At selected fluvial sites it is clearly possible to obtain accurate information on the sizes (palaeodischarges) and ages of multiple ancient floods. Can such data be structured in an appropriate manner for flood-frequency analysis? One way of answering this question is to assert a goal of reconstructing a complete catalogue of palaeodischarges exceeding or not exceeding various censoring levels

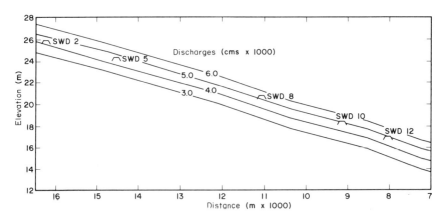

Figure 10.3 Water-surface profiles for indicated discharges along a reach of the Finke River in central Australia. The profiles were generated by a computer flow model fitted to the geometry of the bedrock channel that was measured in the field (see Baker *et al.*, 1983). The heights of numbered slackwater deposits (SWD) are compared to the profiles in order to estimate palaeodischarges.

Magnitude and Frequency of Palaeofloods

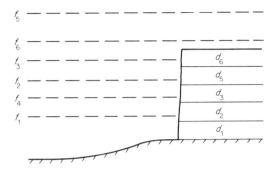

Figure 10.4 Vertical stack of slackwater sediments (d_1–d_6) showing the relationship to a hypothetical sequence of formative flow events (f_1–f_6).

(threshold discharges) over specified time periods. This transforms the question to the field problem of specifying consoring levels that insure a complete catalogue of the largest palaeofloods occurring in a given past time period.

Figures 10.4–10.6 illustrate typical field relationships encountered in SWD–PSI studies. Figure 10.4 shows the 'worst-case' end member of a single, vertically stacked sequence of slackwater deposits. For this case the palaeoflood informational censoring level is the elevation associated with each succeeding deposit, and this level increases with time as new deposits are added to the stack. However, a new flood deposit cannot be emplaced on such a stack unless the responsible flood stage exceeds the elevation of the previous deposit. The censoring level for each flood deposit is the elevation of the top of the deposit emplaced by the previous large flood. Note that this threshold increases when exceedances occur. However, the exact magnitude of the exceedance is unknown, since various depths of flood water above this censoring level are capable of emplacing a new deposit on top of the stack. Note also that floods with stages less than the upwardly moving censoring level are not recorded. For example, in Figure 10.4, floods f_1, f_2, and f_3 are recorded successively in the stack, but flood f_4 is not recorded since it fails to exceed the rising censoring level. Also note that the largest flood f_5 is recorded by a deposit d_5 that underlies the deposit d_6 of a smaller flood f_6. Thus, a single vertical stack of slackwater deposits provides information on minimum palaeostages for the largest floods in a time interval, but it is not exact in specifying those stages nor in the preservation of deposits from smaller floods.

The problems encountered with single-site, vertically stacked slackwater sequences (Figure 10.4) can be minimized by correlation and tracing of flood units. Figure 10.5 shows a more complex slackwater site in which the flood deposits can be traced laterally to their highest levels as they thin upstream along a tributary. An additional check on the maximum flood level is provided by a scour line that can be traced laterally to a dated deposit in a cave. In addition,

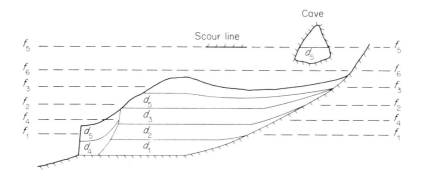

Figure 10.5 Complex slackwater depositional site showing evidence of maximum flood stages and inset deposits, as explained in text. Sequence of floods is the same as in Figure 10.4.

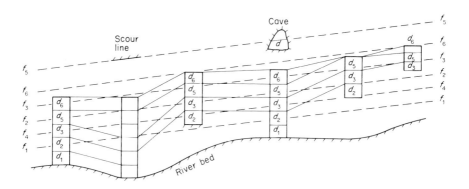

Figure 10.6 Idealized depiction of multiple palaeoflood depositional sites correlated along a study reach. Note the relationships to the formative flood sequence (same as in Figures 10.4 and 10.5) and to the high-water marks for the largest flood.

there is an inset deposit, layers d_4 and d_5, that accumulated below the vertical stack. Note in Figure 5, in contrast to Figure 10.4, that the small flood f_4 is identified and that the magnitude of the largest flood f_5 is apparent. Small floods are identified in the inset because a new, lower censoring level is established. This informational censoring level is not a human artifact, as in the case of historical floods; it is achieved by the natural physical processes of the river system. Similarly the cave provides a very high-level censor of the palaeoflood information. As deposits are correlated from one locality to another the overall palaeoflood record becomes more complete for smaller floods and more accurate for the palaeostages of all floods.

Figure 10.6 illustrates schematically the most general case, involving multiple correlations among multiple sites of the type shown in Figure 10.5. Examples from field situations are given by Kochel *et al.* (1982), Ely and Baker (1985), and

Partridge and Baker (1987). Note that floods failing to leave deposits at one site, because of failure to exceed a local censoring level, may be preserved at other sites. Sites posing a particularly high censoring level for a given flood will preserve the highest deposits emplaced by that flood. Palaeostages for the largest palaeofloods can also be established accurately from scour lines and other direct high water indicators.

Techniques for stratigraphic correlation include geochronology (synchronous flood dates), weathering criteria (palaeosols, induration, colour), intercalated deposits (colluvium, aeolian material), lithology, mineralogy, grain sizes, stratigraphic position, thickness, and sedimentary structures. Catalogue completeness therefore is achieved by (1) lateral tracing of flood deposits to their highest elevation, (2) correlation among multiple sites, (3) evidence of limiting highwater levels, and (4) inset stratigraphic relationships.

FREQUENCY ANALYSIS

Hydrologists and engineers have generally treated the evaluation of magnitude and frequency of flooding as a problem in applied statistics. Their goal, as described by Potter (1987), is the '...statistical estimation of flood quantiles (discharges with specified probabilities of exceedance) for use in water-resource decision-making (e.g. floodplain zoning or design of engineering works).' The usual procedure involves extrapolation from large data sets documenting observed small floods to the quantiles of very large, rare floods. Because of its emphasis on direct, physical evidence of the largest ancient floods, it has only been recently (Stedinger and Baker, 1987) that palaeoflood hydrology has entered the main thread of research into flood-frequency analysis (FFA).

It has recently been recognized that palaeoflood data structured by censoring levels results in a type of systematic data set amenable to FFA. Two types of censoring models are well described: (1) event-based censoring assumes that a discharge estimate for the largest flood in x years is available, that this discharge is inaccurately known, and that no information is available on smaller floods (Hosking and Wallis, 1986), (2) fixed-threshold censoring assumes that all floods exceeding a censoring level (discharge) in time period T are known (Stedinger and Cohn, 1986). Type 1 censoring is applicable to some types of palaeoflood records, such as SS–V–SP data. Type 2 censoring is most appropriate for SWD–PSI data.

Maximum likelihood estimates (MLEs) are used to determine the parameters of a distribution from which floods are assumed to come. This is accomplished by maximizing the probability of observed flood events. Moreover, in contrast to the weighted moment techniques commonly employed in FFA, MLEs are especially versatile in allowing censored unconventional data sets, such as SWD–PSI data, to be combined with conventional systematic data in an FFA

(Leese, 1973; Condie and Lee, 1982; Stedinger and Baker, 1987). Cohn (1986) and Stedinger et al. (1988) describe the mechanics of the MLE approach to FFA.

Hosking and Wallis, using assumption (1) above, concluded: '...palaeological information does not add significantly to the benefits conveyed by regionalization.' Stedinger and Cohn (1986), using assumption (2) above, found that threshold-exceedance MLEs employing floods exceeding a known censoring level are nearly as accurate as MLEs employing the actual flood magnitudes. They also found that in such analyses the accuracy of the flood peak is relatively unimportant in the improvement of the FFA provided by the palaeoflood data. They concluded that with the proper structuring palaeoflood data are of tremendous value in FFA

Figure 10.7 illustrates the results of an MLE-based FFA for data on the Salt River in central Arizona, combining a systematic gauge record with SWD–PSi palaeoflood data interpreted by Partridge and Baker (1987). Plotting positions for the palaeoflood data were based on procedures and logic developed in Hirsch and Stedinger (1987). MLE analyses of these data were performed for simple two-parameter distributions (Normal, N, Lognormal, LN, and Gumbel), which failed even to fit the gauge record, and for three-parameter distributions (Pearson 3, P3, log-Pearson 3, LP3, and Generalized Extreme Value, GEV) (Stedinger et al., 1988). Because all distributions showed a tendency to fit the smaller floods in the series at the expense of description of the largest events, analyses were performed by censoring the magnitudes of floods smaller than

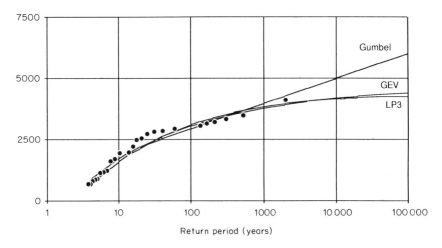

Figure 10.7 Flood-frequency distributions fitted to data on the Salt River in central Arizona. The log-Pearson 3 (LP3) and generalized extreme value (GEV) curves are for palaeoflood and conventional (annual series) data censored at the 70% level (at or above the 3.3-year flood). The Gumbel distribution is for data censored at the 95% level. Units for the ordinate (discharge) are in $m^3 s^{-1}$. (From Stedinger et al., 1988).

Table 10.2 Salt River flood quantiles and their standard errors (S.E.).

Return Period (years)	P3 Distrib. Flood (m^3s^{-1})	S.E.	LP3 Distrib. Flood (m^3s^{-1})	S.E.	GEV Distrib. Flood (m^3s^{-1})	S.E.	Gumbel Distrib. Flood (m^3s^{-1})	S.E.
10	1800	1	1600	2	1700	2		
100	3000	1	3100	2	3000	1	2900	1
1000	3700	1	3800	1	3800	2	4000	3
10000	4300	2	4200	2	4200	2	5000	4

various percentiles of the flood-flow distributions. This seems reasonable since the smaller floods may have no physical relationship to the larger ones (Klemes, 1986).

Table 10.2 shows the results of the analyses by Stedinger *et al.* (1988). It was found that three-parameter distributions (P3, LP3, and GEV) at 70% censoring and the two-parameter Gumbel distribution at 95% censoring all gave reasonably consistent flood quantile estimates for return periods ranging from 10 years to 10 000 years. Since the Salt River data set includes the largest flood experienced in the last 1000 to 2000 years plus the immediately smaller floods in that interval (Partridge and Baker, 1987), it is appropriate to estimate the 1000-year old flood, and to extrapolate with caution to the 10 000-year flood. Stedinger *et al.* (1988) conclude that SWD–PSI palaeoflood data assisted these analyses in two ways: (1) estimating more precisely the parameters of a chosen reasonable parametric model of flood frequency, and (2) distinguishing from among possible flood distributions those that were consistent with the frequency of observed large floods versus those that were not.

This analysis illustrates a change in FFA philosophy posed by the existence of appropriate palaeoflood data. The best analyses of SWD–PSI sites produce accurate estimates of the largest flow events experienced over centuries to millennia on a given river reach. This is information from the upper tail of the parent distribution of floods, not from the many small events near the mean of distribution. Properties of the latter are best studied by gauge records, where such are available. However, where the interest is in rare, great floods, SWD–PSI palaeoflood data are of immense value in flood-frequency analysis.

ACKNOWLEDGEMENTS

The techniques of SWD–PSI palaeoflood hydrology were initially developed with the support of the Division of Earth Sciences, Surficial Processes program, National Science Foundation, NSF Grants EAR 77–23025, EAR 81–19981, and EAR 83–00183. Work in Arizona was supported by the Salt River Valley Water

Users' Association. Discussions with J. R. Stedinger led to improvements in the manuscript.

REFERENCES

Baker, V. R. (1984). Flood sedimentation in bedrock fluvial systems, in E. H. Koster and R. J. Steel, (eds), *The Sedimentology of Gravels and Conglomerates*, Canadian Association of Petroleum Geologists Memoir 10, pp. 87–98.
Baker, V. R. (1987). Paleoflood hydrology and extraordinary flood events, *J. Hydrol.*, **96**, 79–99.
Baker, V. R., and Costa, J. E. (1987). Flood power, in L. Mayer and D. Nash, (eds), *Catastrophic Flooding*, Allen and Unwin, London, pp. 1–24.
Baker, V. R., Kochel, R. C., Patton, P. C., and Pickup, G. (1983). Palaeohydrologic analysis of Holocene flood slackwater sediments, in J. Collinson and J. Lewin, (eds) *Modern and Ancient Fluvial Systems: Sedimentology and Processes*, International Association of Sedimentologists Spec. Publ. No. 6, pp. 229–39.
Baker, V. R., and Pickup, G. (1987). Flood geomorphology of the Katherine Gorge, Northern Territory, Australia, *Geol. Soc. America Bull.*, **98**, 635–646.
Baker, V. R., Pickup, G., and Polach, H. A. (1983). Desert palaeofloods in central Australia, *Nature*, **301**, 502–4.
Baker, V. R., Pickup, G., and Polach, H. A. (1985). Radiocarbon dating of flood deposits, Katherine Gorge, Northern Territory, Australia, *Geology*, **13**, 344–7.
Cohn, T. A. (1986). *Flood Frequency Analysis with Historical Flood Information*, Ph.D. Dissertion, Cornell University, Ithaca, N.Y.
Condie, R., and Lee, K. A. (1982). Flood frequency analysis with historic information, *J. Hydrol.*, **58**, 47–61.
Costa, J. E. (1983). Paleohydraulic reconstruction of flash-flood peaks from boulder deposits in the Colorado Front Range, *Geol. Soc. America Bull.*, **94**, 986–1004.
Ely, L. L., and Baker, V. R. (1985). Reconstructing paleoflood hydrology with slackwater deposits: Verde River, Arizona, *Phys. Geogr.*, **6**, 103–26.
Hirsch, R. M., and Stedinger, J. R. (1987). Plotting positions for historical floods and their precision, *Water Resources Research*, **23**, 715–27.
Hosking, J. R. M., and Wallis, J. R. (1986). Paleoflood hydrology and flood frequency analysis, *Water Resources Research*, **22**, 1606–12.
Hupp, C. R. (1988). Plant ecological aspects of flood geomorphology and paleoflood history, in V. R. Baker, R. C. Kochel and P. C. Patton, (eds), *Flood Geomorphology*, Wiley, N.Y., pp. 335–356.
Klemes, V. (1986). Dilettantism in hydrology: transition or destiny?, *Water Resources Research*, **22**, 177S–188S.
Kochel, R. C., and Baker, V. R. (1982). Paleoflood hydrology, *Science*, **215**, 353–61.
Kochel, R. C., and Baker, V. R. (1988). Palaeoflood analysis using slack water deposits, in V. R. Baker, R. C. Kochel and P. C. Patton, (eds), *Flood Geomorphology*, Wiley, N.Y., pp. 357–376.
Kochel, R. C., Baker, V R., and Patton, P. C. (1982). Palaeohydrology of southwestern Texas, *Water Resources Research*, **18**, 1165–83.
Leese, M. N. (1973). Use of censored data in the estimation of Gumbel distribution parameters for annual maximum flood series, *Water Resources Research*, **9**, 1534–42.
Leopold, L. B., and Langbein, W. B. (1962). The concept of entropy in landscape evolution, *U.S. Geological Survey Professional Paper*, **500-A**, A1–A20.
O'Connor, J. E., Webb, R. H., and Baker, V. R. (1986). Paleohydrology of pool and riffle pattern development, Boulder Creek, Utah, *Geol. Soc. America Bull.*, **97**, 410–20.

Partridge, J., and Baker, V. R. (1987). Paleoflood hydrology of the Salt River, Arizona, *Earth Surface Processes and Landforms*, **12**, 109–25.

Patton, P. C., and Baker, V. R. (1977). Geomorphic response of central Texas stream channels to catastrophic rainfall and runoff, in D. Doehring, (ed.), *Geomorphology of arid and Semi-arid Regions*, Allen and Unwin, London, pp. 189–217.

Potter, K. W. (1987). Research on flood frequency analysis: 1983–1986, *Rev. Geophys.*, **25**, 113–18.

Stedinger, J. R., and Baker, V. R. (1987). Surface water hydrology: historical and paleoflood information, *Rev. Geophys.*, **25**, 119–24.

Stedinger, J. R., and Cohn, T. A. (1986). Flood frequency analysis with historical and paleoflood information, *Water Resources Research*, **22**, 785–93.

Stedinger, J. R., Therival, R., and Baker, V. R. (1988). Flood frequency analysis with historical and paleoflood information, Salt and Verde Rivers, Arizona, *Proceedings of the Eighth Annual Meeting of the U.S. Committee on Large Dams*, Salt River Project, Phoenix, Arizona, pp. 3.1–3.35.

Taylor, R. E., Donahue, D. J., Zabel, T. H., Damon, P. E., and Jull, A. J. T. (1984). Radiocarbon dating by particle accelerators: an archaeological perspective, in J. B. Lambert, (ed.), *Archaeological Chemistry–III*, pp. 333–356, American Chemical Society Advances in Chemistry Series No. 205.

Williams, G. P. (1984). Paleohydrological equations for rivers, in J. E. Costa and P. J. Fleisher, (eds), Development and Applications of Geomorphology, Springer-Verlag, Berlin, pp. 343–67.

Floods: Hydrological, Sedimentological and Geomorphological Implications
Edited by K. Beven and P. Carling
© 1989 John Wiley & Sons Ltd

11 The Use of Soil Information in the Assessment of the Incidence and Magnitude of Historical Flood Events in Upland Britain

R. F. SMITH
J. BOARDMAN
Countryside Research Unit, Brighton Polytechnic, Falmer, Brighton

INTRODUCTION

The sedimentary record of large floods in Britain, where it is present, is usually devoid of datable material. There is also the problem that rapid rates of vegetation succession in Britain lead to loss of visual evidence of major events. Indirect methods must therefore often be used in the attempt to reconstruct flood history. These methods include observations relating to destruction of artefacts such as stone walls and bridges. Such sources may provide evidence in the short term but the probability of repair and problems of dating make them less suitable over longer time scales. Evidence from tree-growth rings may be useful in the short to medium term but can be difficult to interpret where land-use change is a possibility. Evidence from lichens is valuable over periods up to and in excess of 100 years (Milne, 1982; Harvey *et al.*, 1984) but rapid growth rates curtail the longer term applicability of the technique in the British context.

Assessment of soil profile characteristics such as relative maturity, presence of buried profiles, and evidence of truncation may, however, provide useful evidence of the relative age of erosional and depositional features in river valleys. Such work is described here and relates to the valleys of the Porth-Llwyd, near Conway, North Wales, affected by the Llyn Eigiau dam disaster of 1925 and Mosedale Beck, near Keswick in the Lake District, for which there is recently discovered documentary evidence of a cloudburst in the catchment in 1749. The work demonstrates how evidence from field investigations may be useful in an initial assessment of the relative age of flood deposits and other features.

TERRACE SEQUENCES IN UPLAND VALLEYS IN BRITAIN

Terrace fragments are widely present in upland valleys in Britain. Often there is the implicit assumption that such terraces and associated major flood features such as boulder berms are outwash deposits of Dimlington and/or Loch Lomond Stadial age (Sissons, 1982; Rose and Boardman, 1983). This has been a particularly tempting assumption in valleys such as Mosedale and the Porth-Llwyd where evidence for Loch Lomond Stadial glaciation exists. Recent work by Harvey *et al.* (1984) in the Howgill Fells, Cumbria and Robertson-Rintoul (1986) in Glen Feshie, Scotland, has, however, demonstrated the existence of sequences of terraces of Holocene age in upland valleys. This work has indicated the potential for the use of soils in Britain as a relative dating technique (Birkeland, 1984). Such Holocene terrace sequences, together with observations of the effects of contemporary floods (Carling, 1986, 1987), must throw into question any general assumption that all major flood deposits such as boulder berms in valley bottoms are necessarily the product of meltwater discharges.

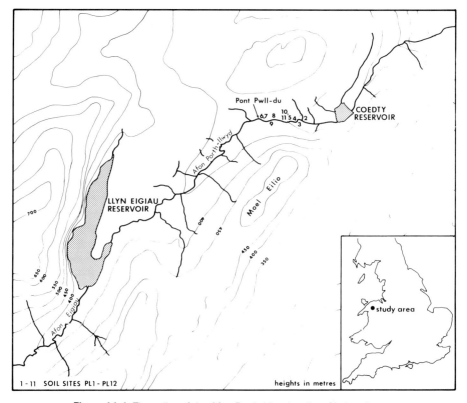

Figure 11.1 The valley of the Afon Porth-Llwyd: soil profile locations.

The investigations described here in the valleys of the Porth-Llwyd and Mosedale Beck show a wide range of soil development on river terraces. It appears that these sequences, like those in the Howgills and Glen Feshie may be representative of episodes of aggradation and incision over Holocene as well as Devensian time. In both valleys, however, a recent flood episode appears to have disrupted the lower elements of the terrace sequence with accompanying modifications of terrace soils and the deposition of fresh sediments on which soil formation has been initiated.

Porth-Llwyd

The Afon Porth-Llwyd (Figure 11.1) flows from Cwm Eigiau through Llyn Eigiau to the Conway at Dolgarrog. The catchment area is 31 km². The solid geology consists predominantly of Ordovician sedimentary sequences and volcanics with much of the catchment being covered by drift derived from slates, shales and grits, acid and basic igneous rocks (Ball, 1963). Rainfall increases rapidly with altitude and is in the region of 1800 mm at the study sites. The altitude of the study sites is of the order of 330 m and the mean annual temperature around 7°C. The soils of the catchment are predominantly brown podzolics of the Manod association, cambic stagnogleys of the Cegin association and cambic stagnohumic gleys of the Wilcocks 1 association (Rudeforth et al., 1984). Some acid hill peat is also present (Ball, 1963). The terrace sequence affected by the flood is within the reach from Pont Pwll-du to the Coedty reservoir. The channel is cut in till in this section and no rock outcrops are present.

Fearnsides and Wilcockson (1928) document the effects of the flood in 1925 caused by the failure of the foundations of the dam wall at Llyn Eigiau. They suggest that the maximum rate of discharge through the channel below Pont Pwll-du may have been of the order of $156 m^3 s^{-1}$ leading to incision which they estimated at 3–3.5 m in this section.

A well developed terrace and associated soil sequence is present in the valley of the Porth-Llwyd (Figure 11.2). A tripartite division of this sequence is appropriate to this study. Terrace remnants between 7 m and 3.5 m above the channel show well-developed brown earth and brown podzolic profiles. Between 3 m and 3.5 m some terrace remnants show evidence of profile truncation and burial by fresh alluvium while a third group of profiles at and below 3 m are developed in fresh alluvium.

Representative profiles (Figure 11.3) illustrate these contrasts. Profile PL2 shows substantial profile development with clearly defined A and B horizons to a depth of over 60 cm. This soil shows podzolic characteristics but some other terrace profiles at and above 4 m were brown earths. Some profiles contain cemented B/C horizons strongly stained by manganese. In contrast soils developed in 1925 flood deposits show only poor horizon differentiation (Profile

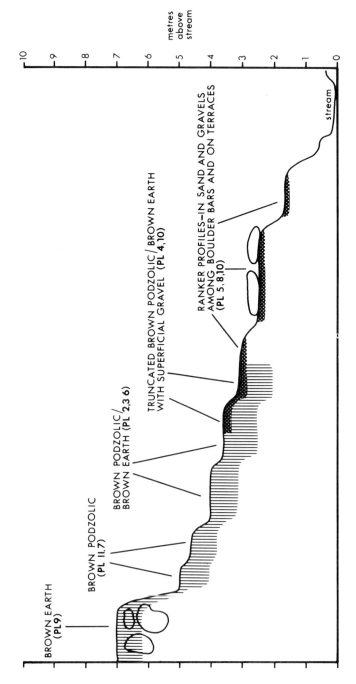

Figure 11.2 Afon Porth-Llwyd: generalized terrace sequence and associated soils.

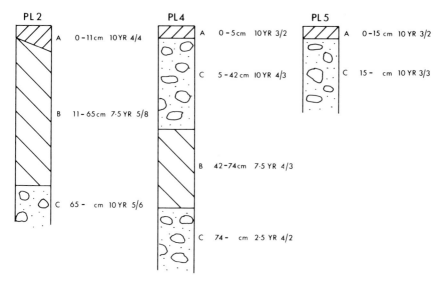

Figure 11.3 Afon Porth-Llwyd: representative soil profiles (for profile locations see Figure 11.2). Soil horizon notation after Hodgson (1976). Munsell Soil Colours are also given.

PL5, Figure 11.3). This profile, developed in coarse sand and gravel, shows a thin A horizon differentiated by a slight colour difference and finer texture from C horizon material below. These differences reflect some organic matter accumulation in the A horizon and a certain amount of differential sorting of finer particles towards the top of the profile, probably resulting from the activities of ants. Profile PL4 shows a similar A/C soil developed in flood deposits but overlying the B and C horizon of a truncated profile. The extent of A horizon development in this profile is comparable to that of Profile PL5.

In the valley of the Porth-Llwyd there is little evidence of truncation of soil profiles without subsequent burial by flood deposits nor, except in one restricted backwater location, were silty, as opposed to sand and gravelly, flood deposits encountered.

Mosedale

Mosedale Beck (Figure 11.4) has a catchment area of 8.1 km². The channel is located predominantly in sandy tills although stream sections confined by slate bedrock occur in places. Two tills are present. The lower, infilling a deep buried valley, is the Thornsgill Till, strongly weathered in its upper section to form the Troutbeck Palaeosol (Boardman, 1985). The upper till, the Threlkeld Till, is the Late Devensian till of the northern Lake District. The Thornsgill Till contributes

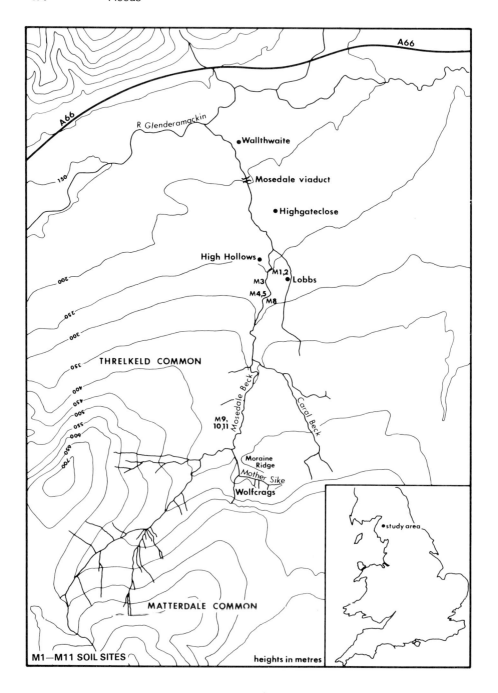

Figure 11.4 The valley of Mosedale Beck: soil profile locations.

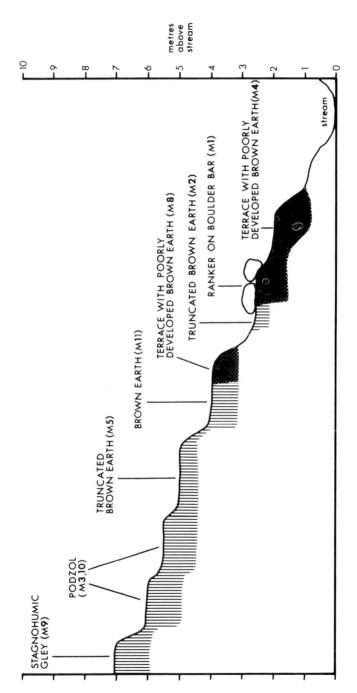

Figure 11.5 Mosedale Beck: generalized terrace sequence and associated soils.

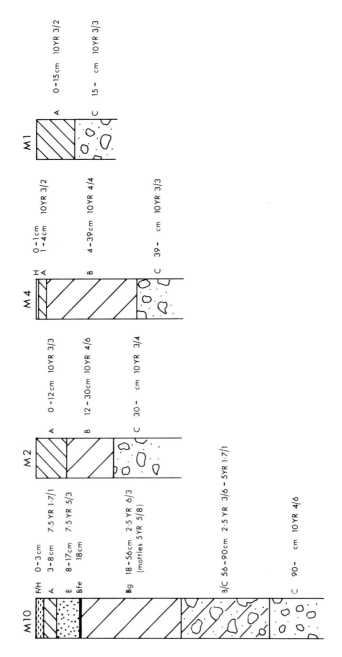

Figure 11.6 Mosedale Beck: representative soil profiles (for profile locations see Figure 11.4).

strongly weathered clasts to the alluvial load of the stream and these are visible in contemporary channel deposits several hundred metres downstream of the major outcrops. The principal soil types of the catchment are cambic stagnohumic gleys of the Wilcocks 1 association, typical brown podzolic soils of the Malvern association, and podzolic soils of the Hexworthy association (Jarvis et al., 1984).

The soil sites studied are at altitudes between 250 and 380 m. Rainfall is of the order of 1700 mm and mean annual temperature is about 7°C (by adjustment of data for Newton Rigg, Penrith—Matthews, 1977). Evidence has been found recently for the occurrence of a major flood event in the valley in August 1749. A letter in *Gentleman's Magazine* 1754 refers to a highly localised cloudburst over the catchment which caused much damage along Mosedale Beck (see Hutchinson, 1794). There is little clear evidence of this today; most stone walls, for example, have been repaired but a leat used for water supply has been partially washed away in the area above the abandoned farm at High Hollows.

The highest terrace at Mosedale is about 7 m above the present stream channel (Figure 11.5). Again a tripartite division is evident with higher terrace fragments showing strongly developed soil profiles to depths in excess of 1 m, terraces of intermediate height showing, in some locations, truncation of the superficial horizons of well-developed soils while lower terraces exhibit poor soil development in gravelly, sandy and silty deposits. Four profiles (Figure 11.6) are indicative of the range of variation. Profile M10 from a terrace at 6 m exhibits podzol characteristics and the development of a strongly cemented sand and gravel horizon at the base of the profile. This horizon is dominated by dark colours indicative of manganese and iron accumulation. Profile M2 represents a truncated brown earth at a height of 2.5 m while profile M1, at a similar height, is a ranker developed on a boulder bar in gravel and coarse sandy deposits. Some young profiles in Mosedale, however, are developed in silty flood deposits (Profile M4, Figure 11.6). Here 39 cm of silty sediment with a thin A horizon overlies a C horizon of coarse sand and gravel infilling between large imbricate boulders up to 70 cm in length. The B horizon of this profile is partially weathered and of uniform appearance throughout. Such a horizon appears to represent deposition of transported 'soil' material rather than substantial *in situ* weathering and pedogenesis.

DISCUSSION

The valleys of the Porth-Llwyd and Mosedale Beck have broadly similar soil-forming environments. There is little reason to suppose that trends in soil development over time will diverge widely. The input of partially weathered material from the Troutbeck Palaeosal in Mosedale is a potential source of difficulty but is unlikely to influence greatly the broad considerations advanced here.

In both valleys soils developed on lower terraces are characterised by shallow A horizons and absent or poorly developed B horizons. Soils developed on coarse sand and gravel deposits on boulder bars show an A/C horizon organisation. In the valley of the Porth Llwyd the boulder bar soils represent about 60 years of soil development since the 1925 flood. In Mosedale the marked contrast in the extent of soil development between the soils of the valley floor, boulders bars and higher terraces indicates that it is unlikely that the terrace sequence in Mosedale is broadly contemporaneous and fits into a Dimlington–Loch Lomond Stadial timescale as has been suggested previously (Rose and Boardman, 1983). The similarity between the boulder bar soils in the Porth-Llwyd and Mosedale suggests that the latter are young and in view of the documentary evidence for a major flood in Mosedale in 1749 it is appropriate to assign these flood deposits to that event. The A horizons of the Mosedale boulder bar profiles are deeper than the Port-Llwyd profiles. This is consistent with the longer time period for soil formation in Mosedale, over 240 as opposed to 60 years.

Most flood deposits in the Porth-Llwyd have a finer fraction composed dominantly of sand and gravel and this is therefore the major soil parent material. In contrast there are a number of locations in Mosedale where silty deposits are present. These too show thin A horizons and other immature features, but are not as clearly visually differentiated from, for example, brown earth soil profiles on higher terraces. In the use of soils to make a field assessment of the relative age of deposits there seems to be least scope for confusion in confining examination to soils of boulder bars and similar deposits where soil parent materials are relatively unweathered sands and gravels. It may be more problematic to interpret profiles which incorporate substantial amounts of partially weathered finer grained materials derived from soil erosion within the catchment. More deposits like this were encountered in Mosedale than in the Port-Llwyd, perhaps because the former flood resulted from a major storm affecting the catchment while most of the Porth-Llwyd sediments were derived from the vicinity of the channel. The work in Mosedale and in the valley of the Porth-Llwyd indicates that soil information may also provide useful evidence relating to the vertical extent of a flood. In the Porth-Llwyd truncated soil profiles are present in several locations at heights of 3 and 3.5 m above stream level but are not seen above that height. There appears to be broad agreement between the soil information relating to flood height and that derived from features such as broken walls, but this will be reported elsewhere.

In Mosedale truncated profiles show greater height variation extending from 5 m to 2.5 m above present stream level. Mosedale also shows greater variability in characteristics such as valley gradient and valley width. This may have resulted in more variation in the height of the flood wave than in the Porth-Llwyd. Also such contrasts may have influenced the depth of incision and extent of sedimentation associated with the flood event. The use of soil information to correlate terrace remnants might provide a series of benchmarks against which to estimate the extent of such incision or deposition. Valley constrictions relating to the rock cut channels present in Mosedale may also have resulted in temporary boulder blockages and therefore variation in flood height. In view of the frequent difficulty of establishing the depth of floods, soil information may thus add an important component to data bases required in palaeohydrologic reconstruction. Evidence of profile truncation and from soil development on high-level boulder bars may also add to our knowledge of the number of flood events in the past.

Soil information may also be of value in the assessment of the lateral extent of flooding in a valley and the extent to which valley side slopes were destabilised as a result of a flood event. In Mosedale, for example, soil investigations in progress suggest that a complex system of anastomosing high-gradient palaeochannels extending to a distance of over 50 m to the east of the present channel may have been formed by the 1749 flood event. Soils on some adjacent valley side landslips appear poorly developed and further work involving inspection of a wider range of soil parent materials in combination with laboratory and statistical analyses might make landform age correlations possible.

A range of analytical techniques are also applicable to studies of comparative development of soil profiles. Organic matter concentrations in A horizons may be useful but appear to reach a steady state more rapidly than any other soil property (Birkeland, 1984). Initial rates of accumulation are often very rapid in the British Isles. Wilson (1987) records over 7% after 31 years in sand dunes in Northern Ireland, which is exceptionally rapid, but Davies and Lewin (1974) show a mean value of 7.2% (loss on ignition) in a floodplain zone dated 1845–68 in Wales. Rates of accumulation slow markedly beyond a few hundred years. Carbonate content and pH may also show trends decreasing with age but in the acid alluvium of many British upland rivers a relative equilibrium may be reached rapidly.

Iron, aluminium and phosphorus may also show trends as soils develop. Davies and Lewin (1974), for example, show a progressive increase in the free iron content of topsoils in successively older zones of the River Rheidol floodplain in Wales. Harvey et al. (1984) also consider iron weathering in alluvial deposits. Birkeland (1984) provides a comprehensive review of a range of analytical techniques appropriate to the examination of profile development.

In using soil information to attempt to reconstruct flood history there is, however, the difficulty that criteria based upon degree of profile development may not differentiate between closely spaced flood events. Such interpretation would require a more detailed knowledge of the chronology of soil development on river terraces than exists at present. There is also the possibility that gaps in the soil record exist. This will depend on the order of high- and low-magnitude flood events. A high-magnitude event which succeeds one of lower magnitude is likely to destroy pre-existing soils.

CONCLUSION

Burke and Birkeland (1979) make a plea for the use of relative dating techniques in combination rather than singly. The potential for using soils along with other approaches including absolute dating has already been demonstrated by Harvey et al. (1984). We would also claim that, within the timescale of historic floods, soil evidence is of value in the elucidation of the flood history of British uplands. Scope exists to apply the methodology to older flood deposits but this will depend upon the establishment of a greater understanding of the relative and absolute chronology of soil development on upland river terraces.

ACKNOWLEDGEMENT

We wish to thank the Countryside Research Unit, Brighton Polytechnic, for financial assistance with the cost of the fieldwork.

REFERENCES

Ball, D. F. (1963). The soils and land use of the district around Bangor and Beaumaris, *Memoirs of the Soil Survey of Great Britain*, H.M.S.O., London.

Birkeland, P. W. (1984). *Soils and Geomorphology*, Oxford University Press.

Boardman, J. (1985). The Troutbeck Paleosol, Cumbria, England, in J. Boardman, (ed.) *Soils and Quaternary Landscape Evolution* Wiley, Chichester.

Burke, R. M., and Birkeland, P. W. (1979). Re-evaluation of multiparameter relative dating techniques and their application to the glacial sequence along the Eastern escarpment of the Sierra Nevada, California, *Quaternary Research*, **11**, 21–51.

Carling, P. A. (1986). The Noon Hill flash floods; July 17th 1983. Hydrological and geomorphological aspects of a major formative event in an upland landscape, *Trans. Inst. Brit. Geogr. N.S.*, **11**, 105–18.

Carling, P. A. (1987). Hydrodynamic interpretation of a boulder berm and associated debris-torrent deposits, *Geomorphology*, **1**, 53–67.

Davies, B. E., and Lewin, J. (1974). Chronosequences in alluvial soils with special reference to historic lead pollution in Cardiganshire, Wales, *Environmental Pollution*, **6**, 49–57.

Fearnsides, W. G., and Wilcockson, W. H. (1928). A topographical study of the flood-swept course of the Porth Llwyd above Dolgarrog, *Geographical Journal*, **72**(5), 401–19.

Harvey, A. M., Alexander, R. W., and James, P. A. (1984). Lichens, soil development and the age of Holocene valley flood landforms: Howgill Fells, Cumbria, *Geografiska Annaler*, **66A**(4), 353–66.

Hodgson, J. M. (1976). *Soil Survey Field Handbook*, Technical Monograph No. 5, Soil Survey of England and Wales, Harpenden, 99pp.

Hutchinson, W. (1794). *The history of the county of Cumberland and places adjacent*, Jollie, Carlisle, pp. 194–7.

Jarvis, R. A., Bendelow, V. C., Bradley, R. I., Carroll, D. M., Furness, R. R., Kilgour, I.

N. L., King, S. J., and Matthews, B. (1984) *Soils and their use in Northern England*, Soil Survey of England and Wales, Bulletin No. 10, Harpenden.

Matthews, B. (1977). Soils in Cumbria, *Soil Survey Record No. 46*, Soil Survey, Harpenden.

Milne, J. A. (1982). River channel changes in the Harthope valley, Northumberland, since 1987, *University of Newcastle upon Tyne, Department of Geography Res. Series* 13, 39p.

Robertson-Rintoul, M. S. E. (1986). A quantitative soil-stratigraphic approach to the correlation and dating of post-glacial river terraces in Glen Feshie, Western Cairngorms, *Earth Surface Processes and Landforms*, **11**, 605–17.

Rose, J., and Boardman, J. (1983). River activity in relation to short-term climatic deterioration, *Quaternary Studies in Poland*, **4**, 189–98.

Rudeforth, C. C., Hartnup, R., Lea, J. W., Thompson, T. R. E., and Wright, P. S. (1984). *Soils and their use in Wales*, Soil Survey of England and Wales, Bulletin No. 11, Harpenden.

Sissons, J. B. (1982). A former ice-dammed lake and associated glacier limits in the Achnasheen area, central Ross-shire, *Trans. Inst. Br. Geogr., N.S.*, **7**, 98–116.

Wilson, P. (1987). Soil formation on coastal beach and dune sands at Magilligan Point Nature Reserve, Co. Londonderry, *Irish Geography*, **20**, 43–50.

12 The Yellow River (County Leitrim Ireland) Flash Flood of June 1986

P. COXON
Department of Geography, Trinity College Dublin

C. E. COXON
Environmental Sciences Unit, Trinity College Dublin

R. H. THORN
School of Science, Sligo Regional Technical College, Sligo

INTRODUCTION

The Carboniferous sandstone and shale uplands of Counties Sligo and Leitrim in northwest Ireland appear to be prone to high magnitude flood events and bog failures. Alexander *et al.* (1985; 1986) have identified the sites of at least 13 peat failures in the uplands to the east of the village of Geevagh in southeast County Sligo/southwest County Leitrim. One of these failures, which occurred in October 1984, resulted in the release of 80 000 m³ of peat and water. The authors concluded that saturation of the peat due to heavy antecedent rainfall (which was unable to drain through the underlying impermeable drift) was the primary causal factor.

The most recent recorded peat failure and flood in the region occurred on the morning of 29 June 1986 in the Yellow River catchment to the east of Lough Allen, Country Leitrim (Figures 12.1 and 12.4) when heavy rainfall triggered two peat slides and gave rise to a major flood event.

This paper describes the peat slides, examines the geomorphological response of the river channel and its tributaries to the flood event and attempts to estimate the return period of the event.

CATCHMENT CHARACTERISTICS

The location of the Yellow River catchment is shown on Figure 12.1. The part of the catchment considered here is the upper catchment downstream to section 1(S1 on Figure 12.4) which has an area of 14.8 km². Figure 12.1 and 12.4 show the location of the catchment and sites noted in the text. The catchment is underlain principally by Upper Carboniferous sandstones and shales (Brandon

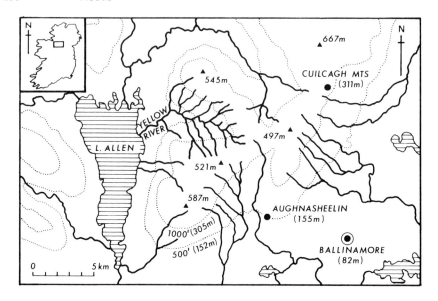

Figure 12.1 Location map of the Yellow River catchment. Circles are the rain gauges referred to in the text. Ireland as inset.

and Hodson, 1984) which are overlain by glacial diamicton. Above 250 m O.D. blanket peat up to 1.5 m in depth is present, while the soils over the rest of the catchment are poorly drained gleys or peaty gleys with silt and clay contents generally in excess of 80% (An Foras Taluntais, 1973). In the upper part of the catchment the channels are deeply incised with steep gradients, while downstream there is a reduction in gradient, greater floodplain development and increased braiding.

METEOROLOGICAL DATA

Thunderstorms are rare in Ireland, being recorded on average 4–7 days each year (Rohan, 1986), but on 27–29 June 1986 severe storms occurred. The main bout of activity began during the evening of Saturday 28 June, and reached a peak at around 1 a.m. on the Sunday morning. Figure 12.2 shows daily rainfall (9 a.m. on 28 June to 9 a.m. on 29 June) in the north midlands of Ireland. The heaviest falls of rain occurred in the Shannon basin north of Banagher, and the highest rainfall was recorded at Aughnasheelin, close to the Yellow River catchment.

The nearest rainfall stations to the catchment are detailed in Table 12.1 (see Figure 12.1 for station locations). Data relating to the storm are shown in Figure

The Yellow River (County Leitrim Ireland) Flash Flood of June 1986

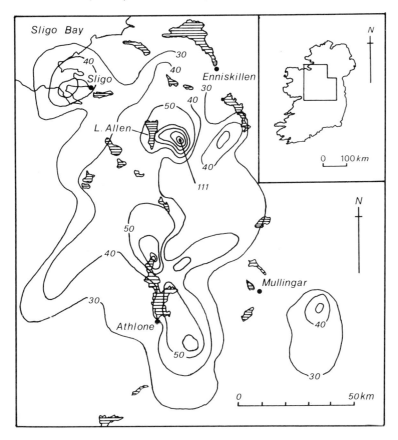

Figure 12.2 Isohyets for rainfall (mm) in the north midlands of Ireland for the 24 h period ending 0900 h on 29 June 1986. The isohyet interval is 10 mm and the 111 mm of precipitation recorded is at Aughnasheelin (see Figures 12.1 and 12.3). The horizontally shaded areas are major lakes. The inset shows the area of Ireland covered by the data.

12.2. The daily rainfall between 9 a.m. on 28 June and 9 a.m. on 29 June was 2.5 times greater at Aughnasheelin, to the SSE of the catchment, than at Cuilcagh Mountains to the ENE, yet the rainfall which gave rise to the flood was clearly heaviest on the ridge to the north-east of the catchment, because the bog slides and the major channel alterations and bedload movement were here. It is likely that very local convectional falls were involved, and presumably the fall recorded 8.2 km to the ENE of the catchment does not represent the rainfall received on the NE side of the catchment itself. A resident in the Yellow River catchment reported that although this area was the worst affected by the storm, the Aughnasheelin area was also badly affected, so the rainfall at the Aughnasheelin gauge (111.0 mm) may be more comparable to the amount that fell in the

Table 12.1 Rainfall stations closest to the Yellow River catchment.

Name	Aughnasheelin	Cuilcagh M	Ballinamore
County	Leitrim	Cavan	Leitrim
Grid ref.	H 085151	H 132248	H 147134
Altitude (m)	155	311	82
Mean annual rainfall (mm)	1312	1845	1141
Station type	Daily gauge	Daily gauge	Autographic recorder
Distance from upper catchment (km)	7.9 (SSE)	8.2 (ENE)	12.9 (SE)

Yellow River catchment. However, the exact amount remains a matter for speculation.

The nearest autographic rainfall recorder, at Ballinamore, received only 38.2 mm during the same 24 hour period; as seen from Table 12.1, this station is lower in altitude and further from the Yellow River catchment. However, it provides evidence of the duration and pattern of the storm. The rain fell in a single event lasting 6.5 hours, from 9.45 p.m. on 28 June to 4.15 a.m. on 29 June, with 84% of the rain falling in the three hours from 1.15 a.m. to 4.15 a.m. The rainfall in the Yellow River catchment may have followed a similar pattern, as local observers report that the flood waters rose slowly initially, and a wall of water appeared at approximately 4 a.m. If 84% of the rainfall at Aughnasheelin also fell in three hours, this would correspond to an intensity of 31 mm h^{-1}; thus the intensity of rainfall leading to the flood may have been of this order of magnitude.

Although the weeks preceding the flood were relatively dry, the main rainfall event on 28/29 June was preceded by a somewhat smaller rainfall event on the 27 June, recorded at both Ballinamore (20.2 mm) and Cuilcagh (20.8 mm), though not significant at Aughnasheelin (3.5 mm), suggesting that this event was also local and convectional in nature. If this preliminary event, 20 h before the main fall, also occurred in the Yellow River catchment, it could have met any soil moisture deficit which might have accumulated during the preceding dry weeks, so that much or all of the rain in the main event could have contributed to runoff.

THE PEAT SLIDES

The locations of the two peat slides, which began at an elevation of about 320 m O.D., are shown on Figure 12.4 (B1 and B2). Both of the slides are characterised by rectangular scars left in the source areas. In both cases crevassing is evident around the edges of the slides and the full depth of peat (*c.* 1.0–1.5 m) has been stripped from the underlying drift. The western flow was

The Yellow River (County Leitrim Ireland) Flash Flood of June 1986 203

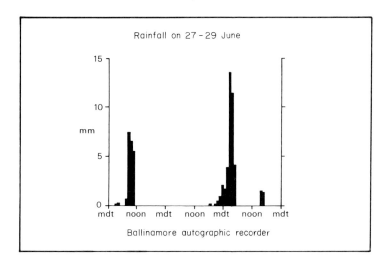

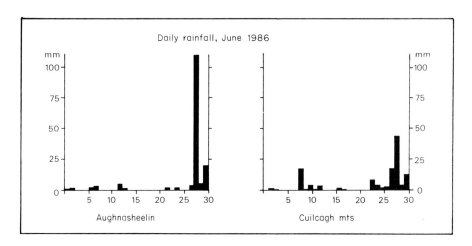

Figure 12.3 Rainfall at Ballinamore (hourly record covering 27–29 June 1986), Aughnasheelin and Cuilcagh Mountains.

about 75 m wide in the upper part of its source area. Bifurcation of the slide resulted in two channels each of about 40 m width. The peat in one of the western slides has been stripped for about 500 m downslope (about 150 m downslope in the smaller slide) and 25 000 m³ of peat are estimated to have been removed and redeposited on the low-lying land near the river channel. The eastern slide was about 20 m wide and peat has been removed for about 350 m downslope. Most of the peat in the eastern flow was discharged into a tributary of the Yellow River.

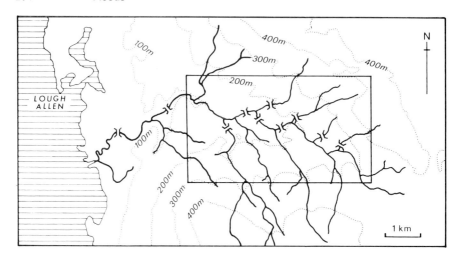

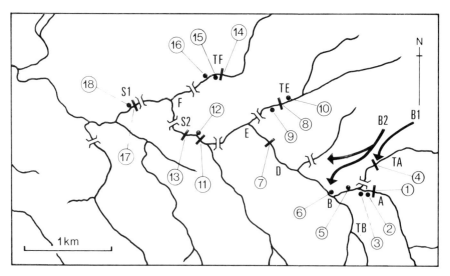

Figure 12.4. The Yellow River catchment, showing the locations of the boulder competence and Manning equations estimates of flow and discharge given in Table 12.2.

In places, up to one metre of the underlying drift has been eroded from the source area of the eastern slide and about 10000 m³ of peat and drift are estimated to have been removed. The peat slides appear to have changed character on the lower slopes and disintegrated into flows containing blocks of peat. The vegetation in the vicinity of the slides is characterised by *Molinia caerulea, Juncus conglomeratus, Polytrichium commune* and *Sphagnum recurvum*. Extensive peat cutting has taken place on the upper slopes and there is a notable convexity in the slope profile in the upper part of each source area with the slope increasing from 6° to 12°.

Alexander et al. (1985, 1986) have described a number of peat failures in the uplands on the western side of Lough Allen and have reviewed the literature on such failures. The slides in the Yellow River area are similar in many respects to those described by Alexander et al. and a similar cause may be invoked, i.e. a combination of site conditions (notably the convexity in slope profile) and a moderate or extreme rainfall event.

Although the peat slides must have released a large amount of water with peat in suspension, there is no evidence of overbank deposition of peat on the floodplain of the Yellow River. The timing of the peat slides with relation to the peak flow in the river is unclear, although large blocks of transported peat were found lying on gravel bars in the Yellow River at Section B, possibly suggesting that the slides added material to the flood waters.

RECONSTRUCTING THE FLOOD

In order to reconstruct the flood, field measurements were made very shortly after the event (where possible) in the Yellow River and its northern tributaries. The geomorphic effects of the event were noted, and these are listed below taking each of the representative sections of the river (shown on Figure 12.4) in turn.

THE GEOMORPHIC EFFECTS OF THE FLOOD

Section A. Numerous shallow slides of the vegetation mat and underlying till occurred (4 m high and 10 m long on average). Where the channel widens and the gradient falls there was widespread deposition and the formation of boulder bars and channel blockages (boulder jams) with steps to 2 to 3 m (Figure 12.5(a)).

Tributary TA. A small tributary (TA), entering in this reach from the northern side of the catchment, carried some of the debris from one of the peat slides as well as runoff generated during the flood. Little geomorphic change appears to have occurred, although a bridge over this tributary was destroyed.

Section B. A fan of debris including some very large boulders was deposited on a wide (> 60 m) meander scar (Figure 12.5(b)). The wide meander curve at B is strewn with peat blocks (up to 1 m in diameter) from the bog slides.

Section D. A meander curve of the Yellow River was cut off during the flood as boulder accumulations formed in the channel, blocking the river and forcing

206 Floods

Figure 12.5 (a) Section A looking upstream at a boulder jam that has produced a step of between 2 and 3 metres in the stream profile. The flow is falling over the step to the right of the photograph. (b). Looking upstream at an accumulation of boulders on the wide section of section B. These boulders were clearly transported and were used in the boulder competence analyses (B/LC/1). (c) The river here at section D flows left to right. The present channel is to the right of the picture and the chute that operated during the flood can be seen below the hay field. Some of the debris that blocked the chute can be seen on the left of the photograph.

some of the stream flow into palaeochannels behind old (possibly 2–300 years?) man-made flood barriers. During peak flow, water occupied a chute (Figure 12.5(c)) across the point-bar of the meander and fell over a number of 2.5 m high waterfalls at the downstream end. These waterfalls began to erode upstream, entrenching the chute system. However, the infilling of the main channel and the production of bars appears to have both raised the main channel and blocked the upstream entrance to the chute, effectively cutting it off from the main flow, with the result that the river remains in the main channel and the chute is unadopted.

Tributary TE. A large volume of coarse sediment was dumped along the channel banks and in boulder bars upstream of a bridge (which had to be reshored due to erosion during the flood). Opposite the boulder bars considerable bank undercutting has produced numerous shallow landslips. Some very large clasts (e.g. 1.7 × 1.5 × 1.2 m) have simply been undercut from the banks of the channel and fallen into the flow. It is hard to judge the distance that such clasts have been transported.

Section 2. In the upstream section, complex deposition led to the formation of a large mid-channel bar with an erosional chute to one side, the main channel the other. Considerable erosion occurred on the opposite bank to the bar (Figure 12.6(a)). Downstream the channel becomes appreciably less steep at 1.5° and the river appears to have deposited large volumes of bedload along

208   Floods

Figure 12.6 (a) The upstream part of section 2 showing the mid-channel bar and erosional chute. (b) The bridge at section 2 that was undercut and subsequently collapsed adding concrete and other debris to the flow. (c) The large landslip (90 m wide) downstream of section F.

this section, producing a wide braided channel. However, this has now been modified by machinery to form a single channel.

Downstream of section 2 is a bridge which was partly destroyed during the flood (Figure 12.6(b)) due to the channel widening and undercutting of the banks and the bridge foundations.

Section F. The river here spilled over its broad floodplain (forming an unadopted chute cutting across a meander) and bank erosion has caused considerable landsliding, including a tripartite rotational slump over 50 m wide and 15 m high. Debris deposited in this reach included iron girders and concrete blocks from the bridge in section 2. One of the latter measured 0.9 × 0.5 × 0.3 m and had been transported over 180 m from the bridge.

Downstream of section F a large, formerly stable, landslip (90 m wide and 40–60 m high) was reactivated by erosion along the channel sides (Figure 12.6(c)). This slide, like the others in the area, supplied a considerable volume of till (which includes the sandstone clasts that form the most common component of the bedload) to the Yellow River.

Tributary TF. This stream joins the main river in this reach. Local residents commented that this stream was greatly altered by the flood and it is clear from eye witness accounts (and the discharge estimates—see below and Table 12.2)

Table 12.2 Mean velocities and discharges estimated using different techniques.

Location number on Figure 12.4	Section	Site number	Techniques (see key below)									
			1		2		3		4		5	
			V	Q	V	Q	V	Q	V	Q	V	Q
	A											
1		A/MN/1	3.75		2.77							
2		A/LC/1	2.85	65.53	2.06	48.41	3.47	60.59	3.82	66.65	3.32	57.9
3		A/LC/2	3.81		3.43							
3		A/LC/2A	3.54		3.71							
3		A/LC/3	3.49		3.42							
5		A/B/LC/1										
	TA											
4		TA/MN/1									3.19	22.77
	B											
6		B/LC/1	4.1		4.24		4.16		4.64			
	D											
7		D/MN/1									1.99	68.63
	TE											
8		TE/MN/1									3.60	39.58

211

	section											
9	TE/LC/1	4.15	45.66	3.81	41.84	4.19	46.02	4.67	51.29			
10	TE/LC/2	4.20	46.20	3.88	42.63	4.26	46.80	4.75	52.22			
2												
11	2/MN/2									3.34	208.91	
12	2/LC/1	3.83	240.20	3.65	228.64	3.89	243.44	4.31	269.95			
12	2/LC/2	5.13	321.36	5.07	317.52	5.13	321.14	5.80	363.11			
12	2/LC/3	4.59	287.56	4.45	278.69	4.59	287.48	5.15	322.53			
13	2/MN/1									2.87	73.80	
TF												
14	TF/MN/1									4.45	100.23	
15	TF/LC/1	3.6	81.93	3.49	78.55	3.76	84.51	4.15	93.48			
16	TF/LC/2(*)	4.47		4.47		4.84		5.45				
1												
17	1/MN/1									3.84	152.57	
18	1/LC/1	4.30	170.85	4.12	163.63	4.30	170.89	4.81	190.85			

(*) indicates a sample measurement of 5 clasts from a fossil berm
Techniques used: 1. Helley (1969) equation (3)
2. Mears (1979), equation (7)
3. Costa (1983), equation (8)
4. Costa (1983), equation (10)
5. Manning Equation, Gray and Wigham (1970) and Cowan (1956)

The first symbol in the 'section' column is the channel reach studied. Prefix MN is a station surveyed for Manning equation analyses. Prefix LC indicates large clast data were collected. The final digit is the sample or section number. The stations measured are shown in Figure 12.4.
V = mean velocity of flow (m s^{-1})
Q = mean discharge (m^3 s^{-1})

that this particular tributary carried a very large volume of water during the flood and that much of the rain that caused the event fell on this northern side of the catchment.

The tributary contains a very large fossil boulder bar over 70 m long and 20 m wide that the flood described here did not affect. Measurements of the clasts in this fossil feature (TF/LC/2 in Table 12.2) show that even larger flows have occurred in this stream in the past, although it is difficult to estimate the size or age of any previous event.

Section 1. The floodplain widens immediately downstream of a bridge, and includes a large, partly vegetated boulder bar, the central core of which does not appear to have been activated in the event we describe but was certainly added to. The bar may be the result of deposition occurring due to channel widening downstream of the bridge. Some of the material added to the bar in section 1 was of boulder size with long axes of c. 0.55-1.25 m and boulders with long axes as large as 1.60 m appear to have been turned over if not actually transported.

VELOCITY AND DISCHARGE ESTIMATES

The peak flood-flow at representative sections of the river was estimated using the Manning equation. Also, the measurement of the long, intermediate and short axes of the five largest boulders that had been transported (confirmed by the overturned areas of sub-aerial weathering and lichen cover and the chipping of the corners of the transported blocks) during the flood at the representative sections, allowed boulder competence calculations to be used. The latter calculations were the same (and were applied in the same way) as those used by Carling (1986). The slope/area method was applied following Gray and Wigham (1970) and standard tables were used to arrive at values of Manning's *n* following Cowan (1956). Some problems were encountered with the slope/area method especially in section 2 where measurements of the peak flow along this section were complicated by the presence of a wide terrace which supported an unknown depth of water during the flood. It is probable that the discharge for this section is an overestimate and that the wide terrace contributed little to the flow.

The techniques listed in Table 12.2 allowed the estimation of the magnitude of the flood by a number of methods, both empirical and theoretical. The similarity of the estimates produced by two independent techniques (boulder competence and the slope/area method) is of interest. However, it should be noted that the catchment is not gauged, and the estimates are therefore approximations that cannot be tested against absolute data.

Carling (1986) comments on the apparent tendency of boulder competence calculations to over-estimate palaeo-discharge. The velocity and discharge estimates obtained in this study (Table 12.2) show little over-estimation compared to values from the slope/area method but bearing in mind the problems associated with the latter technique (e.g. Jarrett, 1987) it may be that the flood estimates arrived at here are high.

THE RETURN PERIOD OF THE EVENT

The return period of the Yellow River flood can be estimated either from the rainfall data or from the estimates of the flood magnitude.

Considering firstly the rainfall data, the initial step is to estimate the magnitude of the event. It is assumed that the depth of rain recorded at Aughnasheelin (110 mm) also fell on the upper Yellow River catchment, and that the rainfall event in the catchment followed a similar pattern to that observed at Ballinamore autographic recorder. This allows for two possibilities, firstly that the rainfall event in the catchment lasted 6.5 hours and secondly 84% of the rain fell in 3 hours (as at Ballinamore), giving a fall of 92.4 mm in 3 hours. This estimate is comfortably within the estimated maximum 3 hour rainfall for the area, of 170 mm, calculated as in Natural Environment Research Council (NERC) (1975).

Logue (1975) estimates the return period of rainfall events in Ireland of various magnitudes and durations. The 3 hour rainfall with a 50 year return period is 39.8 mm, and the 6 hour rainfall with a 50 year return priod is 55.1 mm, so a fall of 92.4 mm in 3 hours, or even the more conservative estimate of 110 mm in 6 hours, has a return period greatly exceeding 50 years. Logue considers that under Irish conditions, no sound basis exists for the extension of the growth curves to return periods greatly exceeding the length of the station records, therefore no values for return periods greater than 50 years are given. However, comparisons can be made with estimated UK return periods for large, rare rainfall events, from various sources, although the validity of such comparisons is questionable. For example, it can be noted that the suggested 92.4 mm in 3 hours is of comparable magnitude to the 87–105 mm in 2.5 hours which gave rise to the Noon Hill floods in the Pennines, for which Carling (1986) quotes a return period of 400–2500 years by the method of NERC (1975), and 200–500 years by Bilham's (1935) method. So the Yellow River flood rainfall return period may be of a similar order of magnitude.

The second approach is to estimate the return period of the flood itself. The UK Flood Studies Report (NERC, 1975) provides two methods of determining flood magnitudes for ungauged catchments. The one used here is the statistical approach, whereby the mean annual flood for the catchment is estimated from the catchment characteristics using a multiple regression equation, and floods of various return periods are then determined using a regional growth curve, which relates the mean annual flood to floods of other return periods. The second approach, the unit hydrograph method, in which the growth curve is derived from rainfall records rather than river flow data, was not applied here because of the lack of reliable data for rainfall return periods greater than 50 years, discussed above.

The first step in the statistical approach is the estimation of the mean annual flood from the catchment characteristics. The multiple regression equation given for Ireland (which is treated as a single region in the Flood Studies Report) is as follows:

$$\bar{Q} = 0.0172 \, AREA^{0.94} \, STMFRQ^{0.27} \, S1085^{0.16} \, SOIL^{1.23} \, RSMD^{1.03} (1+LAKE)^{-0.85}.$$

See NERC (1975) for the meaning of the different variables; estimated values for the Yellow River catchment are: $AREA = 14.8$, $STMFRQ = 1.5-3.0$, $S1085 = 30-50$, $SOIL = 0.45-0.50$, $RSMD = 56.4-59.7$, $LAKE = O$. This gives a mean annual flood estimate of the order of $10-15 \, m^3 s^{-1}$, corresponding to $0.7-1 \, m^3 s^{-1} km^{-2}$, a flow per unit area comparable to mean annual floods in Irish upland catchments included in NERC (1975). Various alternative regression equations for the calculation of mean annual floods in small catchments ($< 20 \, km^2$) have been proposed; Wilson (1983) provides three, from Poots and Cochrane (1979), Poots (1979) and Institute of Hydrology (1978), which use three or four of the NERC catchment characteristic variables. Using these equations, values of $\bar{Q}$ for the Yellow River catchment range from $12.4 \, m^3 s^{-1}$ to $18.8 \, m^3 s^{-1}$, i.e. slightly higher than the values from the National Environment Research Council (1975) regional equation. However, the highest values come from the Institute of Hydrology (1978) small catchment equations, which are reported to be rather less precise, in terms of standard error, than the six variable regional equation.

The second step is the estimation of floods of other return periods from the mean annual flood. The regional growth curve for Ireland relating the flood of a given return period, $Q(T)$, to the mean annual flood, $\bar{Q}$, provided in NERC (1975), extends up to a return period of 200 years. $Q(200)/\bar{Q}$ is 2.14, so the 200 year flood for the Yellow River would be estimated as $21-32 \, m^3 s^{-1}$). Therefore the estimated observed peak discharge (c. $150-190 \, m^3 s^{-1}$) would appear to have a return period of much greater than 200 years.

The fact that Ireland is treated as a single region in compiling the growth curve means that the curve probably largely reflects the behaviour of large, lowland rivers, since these constitute the majority of sites included, and it may not be applicable to small upland catchments. The Institute of Hydrology (1978) suggests that the growth curve for small catchments is not significantly different from the general Great Britain growth curve given in NERC (1975), which extends up to 1000 years. Applying this general curve (on which $Q(1000)/\bar{Q}$ is 4.38) to the Yellow River gives an estimated 1000 year flood of $44-66 \, m^3 s^{-1}$, which is still considerably smaller than the estimated peak flow on 29 June 1986.

However, there is evidence that some of the British growth curves may underestimate $Q(T)/\bar{Q}$ for rare events, and Stephens and Lynn (1978) suggest that the curves ought to be steeper, for prediction purposes, than they are at present. There are 16 instances from the UK of floods with a peak discharge more than five times the mean annual flood (NERC, 1975). For example, the Lynmouth floods of 1952 produced a flood peak in the West Lyn estimated at 221–252 m^{3}s^{-1} (9.4–10.7 m^{3}s^{-1}km^{-2}: a comparable runoff per unit area to the Yellow River Flood), which was about 15 times the estimated mean annual flood.

To summarise, if it is accepted that the mean annual flood in the Yellow River is of the order of 10–15 m^{3}s^{-1}, then the estimated peak flow of the June 1986 flood is an order of magnitude greater, i.e. $Q(T)/\bar{Q} = c.$ 10. From the growth curves given in NERC (1975), this would correspond to a return period of several thousand years. However, the applicability of these growth curves is questionable. All that can be concluded is that the return period is likely to be at least of the order of hundreds of years, which is in agreement with the tentative comments on the return period of the rainfall event. It is interesting to note that the estimated flood discharge of 150–190 m^{3}s^{-1} lies below, but approaches, the estimated maximum flood for the catchment, of 187–224 m^{3}s^{-1}, calculated from the equation of Farquharson et al. (1975).

CONCLUSIONS

The description above shows that the Yellow River flood was a high-magnitude event with a return period possibly measured in hundreds if not thousands of years. Estimating the scale of the event relies heavily upon palaeo-flood analyses, which in turn rely upon both empirically and theoretically derived equations that contain inherent problems. Some boulder competence equations were shown by Carling (1986) to overestimate flood discharges, but in the case presented above the slope/area method and boulder competence methods produce comparable results, suggesting that the estimates of flood velocity and discharge may be credible.

The main geomorphological effects of the Yellow River flood includes bank erosion and landsliding, bar production, bar migration and chute formation. To date, none of the chutes have been adopted as permanent channels, although they appear to have been utilized to a limited extent during post-flood high-discharge events. In addition the channel was infilled and raised in low gradient or wider reaches, braiding being initiated in places, and boulder jams were formed in the upper reaches producing steps in the profile. The high rainfall initiated peat slides in the upper part of the catchment. The overall geomorphological effects are very similar to those described by Carling (1986) in the north Pennines.

ACKNOWLEDGEMENTS

The authors would like to thank Dr Paul Carling for encouragement and advice and The Irish Meteorological Service and Mr Peter O'Shea for access to meteorological data. We would also like to thank Mr Terry Dunne for making hard copies of photographic slides and Mrs Martha Lyons for drafting the diagrams from originals.

REFERENCES

Alexander, R. W., Coxon, P., and Thorn, R. H. (1985). Bog flows in south-east Sligo and south-west Leitrim, in R. H. Thorn (ed.), *Sligo and West Leitrim. Irish Association for Quaternary Studies Field Guide No. 8 (1985)*, IQUA, Dublin, pp. 58–76.

Alexander, R. W., Coxon, P., and Thorn, R. H. (1986). A bog flow at Straduff Townland, Co. Sligo, *Proceedings of the Royal Irish Academy*, **86**, 107–19.

An Foras Taluntais (1973). *County Leitrim Resource Survey, Part 1–land use potential (soils, grazing capacity and forestry)*, An Foras Taluntais, Dublin, 110 pp.

Bilham, E. G. (1935). Classification of heavy falls of rain in short periods, *British Rainfall*, 262–80.

Brandon, A., and Hodson, F. (1984). *The stratigraphy and palaeontology of the Late Visean and Early Namurian rocks of north-east Connaught*, Geological Survey of Ireland Special Paper Number 6, Geological Survey of Ireland, Dublin, 54 pp.

Carling, P. A. (1986). The Noon Hill flash floods; July 17th 1983. Hydrological and geomorphological aspects of a major formative event in an upland landscape, *Transactions of the Institute of British Geographers*, N.S. **11**, 105–18.

Costa, J. E. (1983). Paleohydraulic reconstruction of flash-flood peaks from boulder deposits in the Colorado Front Range, *Geological Society of America Bulletin*, **94**, 986–1004.

Cowan, W. L. (1956). Estimating hydraulic roughness coefficients, *Agricultural Engineering*, **37**(7), 473–5.

Farquharson, F. A. K., Lowing, M. J., and Sutcliffe, J. V. (1975). Some aspects of design flood estimation, *BNCOLD Symposium on Inspection, Operation and Improvement of Existing Dams*, Newcastle University Paper 4.7.

Gray, D. M., and Wigham, J. M. (1970). Peak flow-rainfall events, in D. M. Gray, (ed.), *Handbook on the Principles of Hydrology*, Section VIII, Port Washington, N.Y. Water Information Center Inc.

Helley, J. E. (1969). Field measurement of the initiation of large bed particle motion in Blue Creek near Klamath, California, *U.S. Geological Survey Prof. Pap.* 562-G.

Institute of Hydrology (1978). *Flood prediction for small catchments. Flood Studies Supplementary Report No. 6*, Institute of Hydrology, Wallingford, Oxon., 5 pp.

Jarrett, R. D. (1987). Errors in slope-area computations of peak discharges in mountain streams, *Journal of Hydrology*, **96**, 53–7.

Logue, J. J. (1975). *Extreme Rainfalls in Ireland*, Meteorological Service Technical Note No. 40, Meteorological Service, Dublin, 24 pp.

Mears, A. I. (1979). Flooding and sediment transport in a small alpine drainage basin in Colorado, *Geology*, **7**, 53–7.

Natural Environment Research Council (1975). *Flood Studies Report*, HMSO, London.

Poots, A. D. (1979). *A flood prediction study for small rural catchments*, Unpublished M.Sc. thesis, Queen's University, Belfast.
Poots, A. D., and Cochrane, S. R. (1979). Design flood estimation for bridges, culverts and channel improvement works on small rural catchments, *Proceedings of Institute of Civil Engineers,* **66**, TN 229, 663-6.
Rohan, P. K. (1986). *The Climate of Ireland* (2nd edn), Meteorological Service, Dublin.
Stephens, M. J., and Lynn, P. P. (1978). *Regional Growth Curves*, Institute of Hydrology Report No. 52, Wallingford, Oxon.
Wilson, E. M. (1983). *Engineering Hydrology (3rd edn)*, Macmillan, London.

13 River Channel Changes in Response to Flooding in the Upper River Dee Catchment, Aberdeenshire, over the Last 200 Years

LINDSEY J. McEWEN
Department of Geography and Geology, The College of St Paul and St Mary, The Park, Cheltenham

INTRODUCTION

Research into river channel adjustment in upland Scotland has focused on two contrasting timescales. Studies of Late-glacial channel adjustment and palaeohydrology provide insight into fluvial processes over longer timescales (thousands of years; e.g. Maizels, 1983). Over short timescales (1–30 years), individual case studies, which assess the geomorphic impact of a major flood (e.g. Acreman, 1983) or the monitoring of process-response at a particular stie (e.g. Werritty, 1984), have enhanced our knowledge of current rates of river planform change. There has, however, been little evaluation of channel change in response to floods of different magnitudes and frequencies over the past 100–200 years although it appears unlikely that fluvial processes reworking the valley floors have operated at constant rates throughout that period. Hickin (1983) stresses the importance of studying river processes over an intermediate timespan. It is therefore difficult to evaluate the extent to which present fluvial process rates can be extrapolated back into the past century or the spatial variation in river channel adjustment.

As present rates of channel adjustment can only be considered as a legacy of the total range of a channel's behaviour, it is important to place present process-response into a longer temporal dimension. While the immediate geomorphic impact and persistence of landforms generated by extreme flooding have been studied in many different environments (e.g. Costa, 1974; Patton and Baker, 1977; Schick, 1974; Tricart, 1961) and in several British case-studies (e.g. Anderson and Calver, 1977; Carling, 1986), work within Scotland has been more limited. Werritty and Ferguson (1980), however, reconstructed the cyclical history of channel disruption and recovery in response to floods of varying magnitude on a braided reach on the gravel-bed River Feshie, in the Cairngorms, Scotland over 1, 30 and 200 years, while McEwen and Werritty (1988) studied the immediate geomorphological impact and long-term persistence of an extreme flash flood which occurred in August 1978 in the Cairngorms. Milne (1982) performed a similar exercise on a low sinuosity reach within the Harthope valley, Northumberland. All these studies support the importance of extreme floods as landforming agents within high-energy Scottish upland environments with high entertainment thresholds.

Study of the spatial variation in channel planform adjustment to flooding has also been limited in the United Kingdom. Hooke (1977) has, however, used map sources to monitor planform changes down Devon rivers although no attempt was made in that paper to relate these historical changes to stresses of a particular magnitude. There is therefore a lack of information about the extent to which planform change can vary within a region in response to the magnitude and frequency of the flood history.

This study focuses on river planform adjustment in response to flooding in an upland basin, the upper River Dee above Crathie in Aberdeenshire, Scotland over the last 200 years. This work is part of a larger study which assesses the range of planform types, the nature of fluvial controls and the associated modes and rates of planform adjustment (see McEwen, 1986). The following questions are considered:

(1) What is the planform response to runoff events of varying magnitude and frequency?
(2) Under what circumstances are rare floods geomorphologically significant?
(3) What impact may climatic fluctuations and land-use changes have had on rates of planform response, through changes in hydrological regime and sediment availability over the last 200 years?

Such analyses entail the reconstruction of both the history of channel pattern change and the hydrological record within the study area. This paper briefly outlines the map-based approach adopted to assess channel adjustment and summarises the main results. The reconstructed history of flooding and land-use changes within the catchment are presented. Changes in channel pattern are interpreted in relation to the possible impact of random high-magnitude floods, climatic fluctuations and land-use changes.

ENVIRONMENTAL SETTING

The River Dee above Crathie, Aberdeenshire, draining $573 \, km^2$, comprises the steep south-facing slopes of the Cairngorm massif and also the north-facing slopes of the Grampians (Figure 13.1). The solid geology is dominantly Moinian

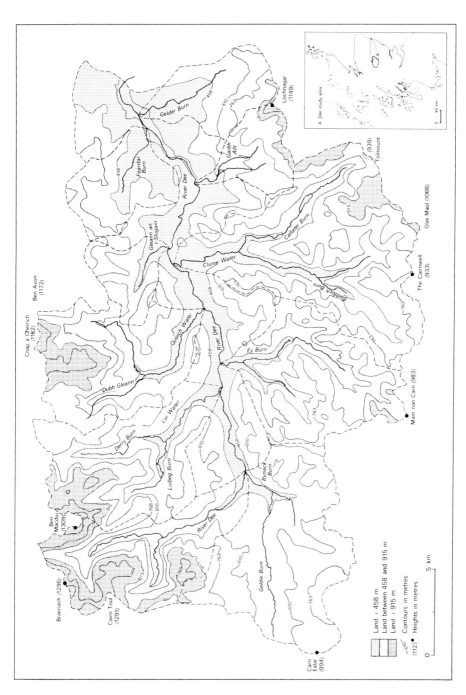

Figure 13.1 General location of the study area.

schist but in the headwaters of the Quoich, Lui and Dee catchments, there are granitic intrusions which form the highest peaks (e.g. Ben MacDhui, 1309 m O.D.).

The glacial legacy of the catchment is very important in determining the spatial variation in fluvial controls. Whereas the upper parts of the catchments are dominated by erosional landforms, such as steep-sided glacial troughs (e.g. Glen Geusachan), the middle to lower reaches are principally depositional zones (see Sissons and Walker, 1974; Sissons, 1976). Frequently within the middle reaches where the valleys widen, there are undulating fluvioglacial deposits to be reworked (e.g. along the middle Lui Burn). Near the confluences with the mainstream River Dee, meltwater gorges exist at a variety of scales, channelling the flow through narrow rock-cut sections (e.g. Linn of Quoich). The lowest reaches comprise areas of fluvioglacial outwash gravels and low-angle gravel fans (e.g. the Ey Burn fan).

In terms of climatic conditions, the catchment is to the east of the area of highest mean annual precipitation in Scotland, with an average annual rainfall of 1164 mm (National Environment Research Council, 1975).

SOURCES

The most complete data-bases available for recording the nature of channel planform are the first and second edition Ordnance Survey 1:10560 (surveyed in 1869 and 1902 respectively) and the metric edition O.S. 1:10000 edition (surveyed 1971) map series. It is, however, possible to extend the record back further by studying channel planform as depicted on William Roy's *Military Survey of Scotland* (1747-55) at a scale of 1:36000 (British Library Press Mark; Maps C.9.B). The detail depicted on these maps has an acceptable level of reliability, especially on mainstream rivers (Arrowsmith, 1809 in Skelton, 1967; Moir, 1973) and checks made with old estate plans confirmed this accuracy. Qualitative distinctions can be made between straight, sinuous and divided channels for the period 1747-55 for comparison with the first edition O.S. map of 1869.

General problems with using maps as geomorphological sources have been outlined before (Carr, 1962; Hooke and Kain, 1982) but there are important points particular to fluvial research. The minimum thresholds of map accuracy (see Harley, 1975) must be carefully considered in interpreting any differences in channel planform between map editions. Arguably, it is the larger changes above this threshold level that will be geomorphologically significant within a macro-scale study. Map information was extensively sustantiated by both field survey and aerial photograph analysis and additional observations are incorporated within the discussion.

Map data can only provide a set of 'stills'; there is no indication if any change

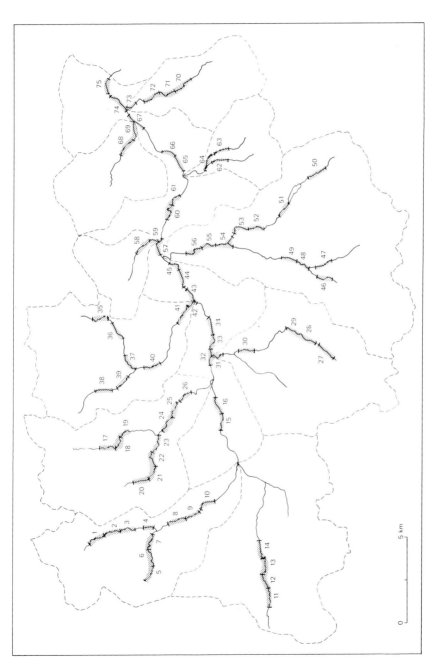

Figure 13.2 Sampled channel segments.

recorded occurred progressively over the intervening period or catastrophically in a few hours but additional contemporary sources can assist in interpretation. Another unknown is the river stage at the time of surveying. Map-based data does, however, provide valuable detailed information, allowing an insight into channel pattern change within upland Scotland, which cannot be obtained from any other source.

SAMPLING PROCEDURE

The population defined for data collection purposes was the river channel depicted as a double blue line on the O.S. 1:25000 edition. The sampling strategy was a simple random sample of 75 1 km segments of river channel (Figure 13.2), which represents approximately 60% of the total stream length. For each of the selected channel segments, a braiding index (BI; adapted from Brice (1960) and Milne (1982)) was measured at each of the three map dates, using a Tektronix digitiser. Description of channel form is based on the classification of Kellerhals et al. (1976), adapted for upland Scotland and map-based information (see McEwen, 1986).

This analysis has two contrasting aims. It is necessary to be able to infer the overall aggregate behaviour of the total population of river channel segments on the basis of the random sample but also to provide analysis of aberrant river sites which may be important geomorphologically. Two forms of analysis of the spatial and temporal variation in planform indices were made; statistical analyses, to test the overall significance of the sample, and visual interpretation of graphical displays to preserve the geomorphic individuality of each contributing sample reach.

Assessing the frequency of braiding index values (Table 13.1), no bars or only localised features (a BI of 1.00–1.09) were indicated on 66–71% of the sample with only 5.4–16.4% having a braiding index > 1.30. On plotting the data on a map base, high values were associated with confluence sites (Figure 13.3(a)). For example, the most extreme value by far (with a BI of 8.80 in 1900) was found upstream of the Quoich confluence with the River Dee where a complex distrib-

Table 13.1 Summary statistics for braiding index.

Parameter	25th Percentile	Median	75th Percentile	Maximum
1867/9	1.00	1.02	1.12	1.51
1902/3	1.00	1.05	1.13	8.80 (1.81)*
1971/2	1.00	1.01	1.11	1.61

*Second largest value

utary fan with a chaotic braided/reticulate pattern occurred (sample 42). Other sites above confluences were also characterised by high values (e.g. the Ey confluence with a BI of 1.68 in 1902; sample 31). Additional reaches with high values include steeper upland confluence sites with major anabranches in the channel (e.g. the upper River Dee with a BI of 1.51 in 1869; sample 4) or where large bars occur on the mainstream River Dee (e.g. on the River Dee above the confluence with the Gleann an t-Slugain with a BI of 1.47 in 1970; sample 57).

Inter-tributary variations were also highlighted; some catchments had very few bars within the sample (e.g. the Clunie Water and Gelder Burn). Possible explanations include channel confinement by terraces or bedrock or lesser availability of fluvioglacial material to be reworked within the channel. In contrast, other catchments had high intra-tributary variation between closely adjoining reaches (e.g. the Lui Water and the upper River Dee; Figure 13.3(a)). This may suggest high variability in local controls, particularly the availability of material and erodibility of banks with periodic sediment storage downstream. Longitudinal profiles down valley showed a series of lower slope alluvial basins interspaced by rock-controlled reaches, which must act as local base-levels (e.g. on the Clunie Water). A complete analysis of the total stream length population would be required for detailed study.

When the median and interquartile range were compared for each map data, a slight increase was noted in 1902 (Table 13.1). However, it is the percentage of values > 1.30 that indicates the main contrast between samples (Figure 13.3(b)). In 1902, 16.4% had braiding indices > 1.30, in comparison to 5.4% and 9.6% for 1869 and 1971 respectively. These changes were verified by palaeochannel evidence and therefore cannot be attributed to variations in mapping practice. Using the Wilcoxon Matched-Pairs Signed-Rank test (Siegel, 1956), a significant difference (0.01 level) was found between braiding index values in 1869 and 1902 and also between values from 1902 and 1971. Comparison of the 1869 and 1971 samples did not, however, yield a significant difference, implying similar lower overall levels of braiding after some disruption during the intervening period. The 1869 and 1971 channel planforms made less use of the available floodplain area for either planform migration or planform expansion.

Analysing the values of individual segments against time, a more detailed pattern emerged (Figure 13.3(c)). It must be noted that 69% of reaches remained relatively stable at all three map-dates while in other areas, more susceptible to change, a variety of different adjustments occurred between 1869 and 1900. High positive increases in the braiding index occurred where samples fell on confluence sites along the River Dee, for example, at the Ey Burn (change in BI from 1.00 to 1.64; sample 31) and Gelder Burn (change in BI from 1.45 to 1.80; sample 73) confluences. This suggests that a threshold for floodplain re-utilisation had been exceeded by 1902 at these sites. The extent of the increase in braiding index was however variable. On the mainstream Dee, for example, the increases in braiding that occurred between 1869 and 1902 were small in

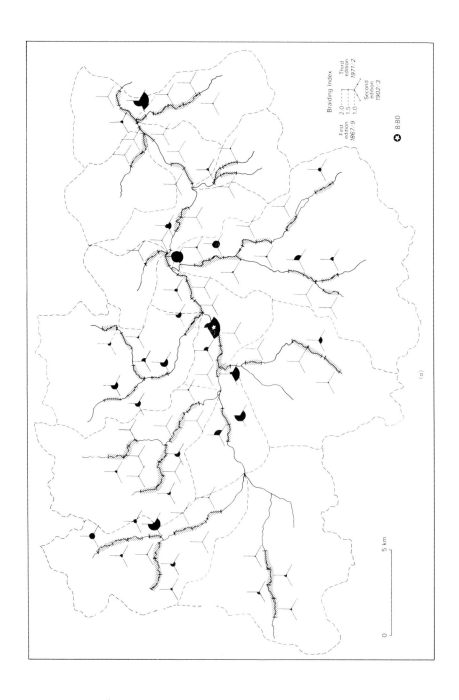

(a)

River Channel Changes in Response to Flooding

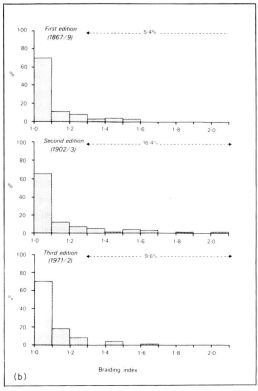

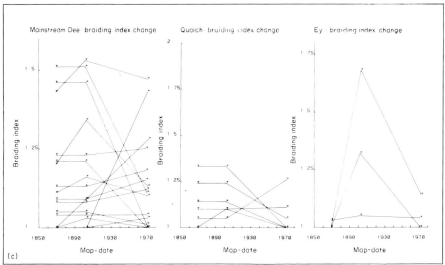

Figure 13.3 Sampled braiding index values: (a) distribution within the study area; (b) frequency distribution; (c) change over time.

comparison to some of the decreases between 1902 and 1971. This implies a long-term adjustment in the system to a pre-1869 period characterized by notable mobility of sediment. The only river that indicated an overall increase in braiding index between 1902 and 1971 was the Clunie Water.

Although between 1902 and 1971 both positive and negative adjustment were recorded, many of the channels whose braiding indices increased between 1869 and 1902 decreased by 1971. This pattern showed in an extreme way in the Ey Burn catchment where several river segments are very close to the threshold between single and divided channels (Figure 13.3(c)).

THE HYDROLOGICAL RECORD

To interpret these planform changes, it is necessary to reconstruct the magnitude and frequency of flooding for the upper Dee catchment over the last 200 years. There are only limited gauged discharge records (with the records on the upper River Dee at Mar Lodge in the study area and Polhollick downstream of Crathie commencing in 1982 and 1976 respectively; Figure 13.4(b)). The extended flood record had to be achieved using a variety of historical sources (e.g. Roberts, 1919; Bremner, 1922; McEwen, 1987(a)) and was substantiated and augmented by analysis of long-term rainfall records (e.g. the Braemar daily record, 1857–1982; McEwen, 1987(b)). The established flood chronology for the River Dee above Crathie (Figures 13.4(a)) allows several observations to be made as to the incidence of fluctuations in flood frequency and the occurrence of apparently random high-magnitude floods.

The most extreme flood on record occurred on 4 August 1829 associated with a summer frontal storm of high intensity and major regional extent, particularly affecting the north to south flowing tributaries (Lauder, 1830). Recurrence intervals for the peak discharge further downstream on the River Dee at Woodend have been placed at as high as 1000 years (National Environment Research Council, 1975). In the 1870s and 1880s, there was an increased frequency of moderate to extreme rainfall peaks (24 hour rainfall > 25.4 mm; see Figure 13.4(c)), accompanied by a climatic fluctuation involving increased annual rainfall totals (McEwen, 1987b). A corresponding increase in flood frequency was found within the reconstructed discharge series, with an increase in moderate to extreme flooding (estimated recurrence interval of 20–50 years) over a similar period. The 1850–1900 period also contained several severe winters coincident with the latter stages of the Little Ice Age so that associated flooding had a substantial snowmelt contribution (e.g. in February 1865).

Since 1902, there have been minor fluctuations indicated in the hydrological record, with the most pronounced being a trough associated with a low frequency of rainfall extremes and floods in the 1970s. Random high-magnitude rainfall peaks generating high-recurrence interval discharges were also evident post-1900. In contrast to the discharge record pre-1900, these major floods were generated by winter cyclonic storms of much longer duration (> 48 hours) in conjunction with exceptionally wet antecedent conditions (e.g. the January 1937 flooding which ranked second after the August 1829 flood). There have been no extreme floods generated by regional summer frontal storms (which rank highly in the extended record) on the upper River Dee above Crathie this century.

In addition to the regional flooding, there was also flooding induced by localised high-recurrence-interval convective storms, which can generate high runoff rates within the tributary catchments (e.g. the August 1956 flooding in the Lui Water catchment; Baird and Lewis, 1957). These events will have little hydrological impact on the mainstream River Dee. Although the frequency of these storms within individual catchments is low, frequency increases when considering the whole upper Dee catchment (see McEwen and Werrity, 1988).

LAND-USE CHANGES

It is also necessary to establish whether any land-use changes have taken place post-1750 which, through alterations to the speed of storm runoff and the shape of the unit hydrograph, could have altered the periodicity of extreme flooding and sediment mobility. Alternatively, the change in land-use since that date may have been hydrologically and geomorphologically insignificant in relative terms to that which took place before 1750. Extensive bioclimatic lowering of the tree-line is evident from the large number of tree stumps buried in the hill peats and was initiated at the beginning of the Atlantic period (5500 B.P.; Jousley, 1973). In the eighteenth and nineteenth centuries, there was large-scale commercial timber felling of pines (*Pinus sylvestris*) on the lower slopes within several tributaries (upper Dee, Quoich and Lui catchments; Pears, 1968; Steven and Carlisle, 1959). Other catchments (e.g. the Gelder associated with Ballochbuie Forest) did not undergo any commercial deforestation (Figure 13.5). No total deforestation of the tributary catchments, however, occurred and there is little quantitative information about the possible hydrological implications of these changes although it is likely there was some change to the speed of runoff. In addition, there must be a threshold in rainfall magnitude beyond which land-use has little impact on catchment response. The deforestation of floodplains leading to increased erodibility during floods is likely to constitute a more important change in planform controls.

Although the study area was selected because human alteration to the natural channel appeared relatively limited, it is important to note that in the mid-eighteenth century, mill-leads associated with the timber industry frequently divided flow along the lower reaches below the meltwater gorges (e.g. at the mouth of the Lui and Quoich catchments; Scottish Record Office, R.H.P. 31322). This would lead to a reduction of flood flows and associated competence

230   Floods

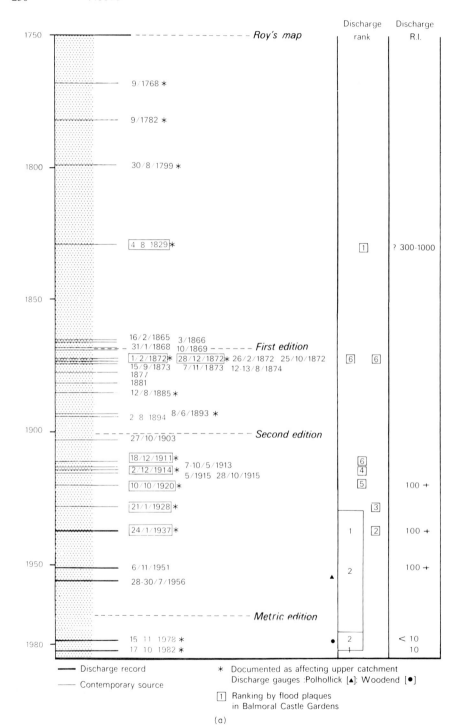

(a)

River Channel Changes in Response to Flooding

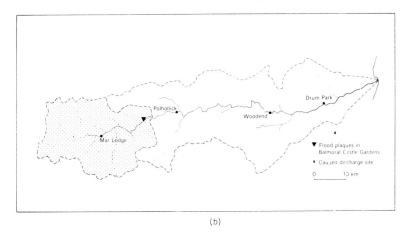

Figure 13.4 (a) Reconstructed historic flood chronology for the River Dee above Crathie (modified from McEwen, 1987b); (b) location of continuous discharge gauges on the River Dee; (c) frequency of rainfall peaks over threshold at Braemar 1857–1982 (Source: McEwen, 1987b).

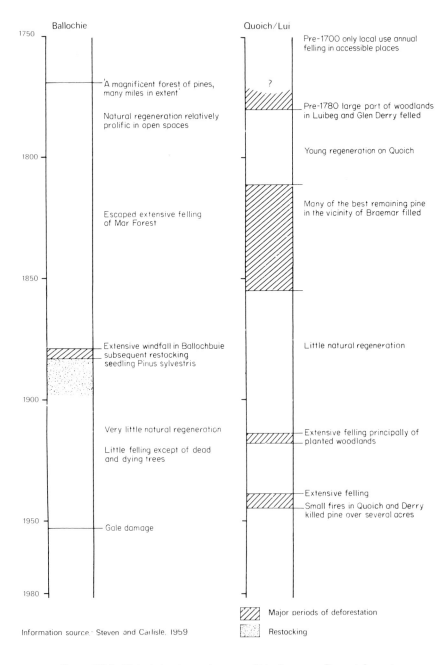

Figure 13.5 Historic land-use changes within the upper Dee catchment.

in the natural channel. In addition, localised embanking of the mainstream Dee, as a remedial measure to prevent bank erosion or inundation of agricultural land, provided artificial channel confinement.

DISCUSSION

Having analysed the macro-scale rates of river planform change, it is now necessary to assess the magnitudes and frequencies of the discharges that are the dominant agents shaping the alluvial landforms on the valley floors within the study area. With the knowledge of fluctuations in the magnitude and frequency of climatic inputs, the occurrence of random extreme floods and the timing and extent of major land-use change, the relative effects of each can be evaluated.

In terms of the impact of climatic fluctuations, the increase in the number of moderate to extreme floods in the 1870s coincides with the period between 1869 and 1902 when there were significant increases in braiding index within the total sample. This could indicate a crossing of a process threshold, a disruption to the fluvial system and a tendency to more divided planforms in unconfined reaches. Several moderate floods in succession would explain the lack of stabilised gravels in a large percentage of the sample, the reworking of islands and the sporadic occurrence of avulsion.

Different catchments varied in the scale of their response to the period of increased flooding; the Clunie Water (lower slopes, more restricted sediment availability) was relatively stable in most sampled reaches with adjustment taking place through minor avulsions. In contrast, the Quoich and Lui Water (steeper slopes, greater sediment availability) underwent several disruptions along their courses.

During the most extreme discharge on record on 4 August 1829, in a few hours there were large sediment inputs to the system and extensive disruption of formerly stable floodplain areas, which completely altered controls on channel planform along change-susceptible reaches (see Lauder, 1830). For example, at the Quoich confluence (sample 42) a major extrinsic threshold was exceeded, leading to extensive planform disruption (cf. the stabler tree-lined planform indicated in William Roy's *Military Survey*, 1747–55), and this new channel alignment has exerted an important control on subsequent channel adjustment. The relaxation time after an event of this extreme magnitude was well in excess of 150 years along some reaches (e.g. sample 42; sample 38). In comparison, the high-recurrence-interval (over 100 years for discharge) winter floods which occrred in 1920 and 1937 (Figure 13.4(a)) had a much lesser geomorphological impact, mainly involved in a reduction rather than an increase in a channel's actively reworked area. After major thresholds have been exceeded, there is a period of lesser response to events of high magnitude until incipient thresholds are re-established (cf. Wolman and Gerson, 1978).

As well as the periodicity of extreme regional storms, which may effect the entire study area, more localised summer convective storms can generate high stream powers in small catchments and can have a lasting geomorphological impact especially where thresholds for sediment entrainment are high. The incidence of these floods may explain discrepancies in the rates of channel adjustment along reaches with similar planform controls in neighbouring tributaries. The flood history within each tributary catchment should therefore be studied. The implication is that neighbouring catchments may be at different distances from quasi-equilibrium depending, not only on their different responses to their individual history of extreme floods, but also on whether these events were regional or localised, generating runoff events of different recurrence intervals in neighbouring catchments. The geomorphological impact of such storms on the mainstream River Dee is, however, negligible.

Despite a similar magnitude–frequency discharge history, adjacent reaches can have more restricted process-responses to major floods due to differences in slope, local sediment limitations or channel confinement by terraces or bedrock. A good example is the distinction between the meltwater gorges and the alluvial fans. Upstream within the steeper gorges, avulsion and other modes of extra-channel adjustment are unable to occur due to channel confinement. With the reduction in constriction downstream on the fans, there is considerable potential for planform disruption. The importance of local base levels in determining the location of reaches actively reworked by floods must also be stressed.

While climatic fluctuation and to a lesser extent random magnitude/frequency variations can cause a regional alteration in channel planform controls, land-use changes are more tributary specific. As major deforestation had already occurred within the Quoich and Lui catchments pre-1850, even the August 1829 flood may have been exacerbated by increased rates of runoff, although the scale of this extreme event would have been largely independent of land-use.

The impact of land-use change on sediment mobility due to increased sediment accessibility after slopes had been deforested is more important. Had the 1829 flood occurred in the early eighteenth century when the middle to lower slopes were more extensively wooded and more stable, its impact might have been less disruptive in terms of accessing new supplies of sediment. It must also be remembered that a catchment's lagged response to such deforestation may take decades to work through the system (cf. Schumm, 1977). It is therefore possible that the increased braiding pre-1900 was partially due to the increased sediment inputs to the channel from this earlier destabilisation of fluvioglacial deposits.

Local modification of the channel planform was occasionally man-induced, explaining disparities between sites where similar patterns of geomorphological response to extreme flooding might be expected. In general, the human impact was localised within the lower reaches of the tributaries. Several mill-leads were destroyed during the 1829 flood and after this time, channels were also readapting to increased flow within a single channel (e.g. above the Quoich and Gelder confluences). This may also have enhanced the geomorphological impact of the subsequent increased frequency of moderate to extreme runoff events in the 1870s and 1880s.

The effects of either climatic or land-use change on planform controls cannot easily be isolated. Any lagged response of the system to deforestation would be co-incident with the climatic fluctuation involving increased flooding. However, the catchments that underwent least deforestation (e.g. the Gelder and Ey catchments) still demonstrated increased channel activity over this period. The inter-relationship between different climatic inputs is similarly complex. The inter-arrival times and sequence of both climatic fluctuations and random high-magnitude events is crucial (cf. Anderson and Calver, 1977). For example, the 1829 flood with its legacy of large sediment inputs provided additional material for the subsequent more moderate floods to rework.

The timescale of inquiry is also important in interpreting channel response to flooding. In paraglacial terms, the fluvial system within the study area is still adjusting over cyclic and graded timescales to the differing glacial legacies (both erosional and depositional) in different parts of the catchments (cf. Church and Ryder, 1972). Present responses may be strongly influenced by both site specific and upstream characteristics which exert controls over much longer periods. For example, meltwater gorges have implications for the flushing of large amounts of sediment downstream although their actual planforms are highly stable. Where planform constraints enforced over longer timescales are spatially relaxed, planform response to extreme flooding is greatest. Different reaches may be adjusting to controls acting over very different time periods, explaining the spatial variation in channel response.

CONCLUSIONS

Planform response to floods of varying magnitudes and frequencies depends on the precise nature of the controlling factors at a given site. Within the upper River Dee catchment, the dominant control is the glacial legacy through channel slope, sediment availability, channel confinement and the positioning of local base-levels. The reach's position in terms of process thresholds, its proximity to a quasi-equilibrium condition and the size of the threshold which must be exceeded for planform change to occur are also highly variable. Frequently process rates do not merely represent a direct response to immediate planform controls due to the often transitory nature of the fluvial system. The legacy of past process-response to floods of different magnitudes over a much longer timespan can be important in assessing the controls on channel planform adjustment over a shorter period (e.g. on the Quoich or Ey fans).

Generalisation is difficult when assessing the circumstances that allow rare floods to be geomorphologically significant but nevertheless certain overall trends can be identified. An extreme event of high recurrence interval (in excess of 100 years) will have a major initially disruptive impact providing that room is available for expansion of the channel's active area and that thresholds for disruption are surpassed (in terms of competence to transport available sediment). It is also important that neither internal nor external thresholds for channel disruption have been recently exceeded as there may be a time-lag before incipient threshold conditions can be attained again. Subsequent more moderate discharges (10-50 years recurrence interval) appear to be more important in returning the channel to a quasi-equilibrium form than disrupting it (Wolman and Gerson, 1978). However, this is not always the case and channel response is highly dependent on both the sequence of floods and planform controls (e.g. the size of the bed material and the channel slope).

Both climatic fluctuations and land-use changes have occurred over the 200 year timespan, which are likely to have affected the magnitude and frequency of flooding and sediment mobility, explaining variations in planform response over time.

It must finally be appreciated that there are limits to understanding in this kind of enquiry as history cannot be completely reconstructed or the past scientifically controlled. Nevertheless, there is no other method which can provide insight into the spatial and temporal variations in planform response to flooding in upland Scotland over an intermediate timescale. It should be emphasised that this form of study, as well as identifying the more active reaches within a region, represents an essential background to further field-based investigation. It is only by this wide-based and integrated approach that the analysis of present-day flood response can begin to be incorporated into the broader study of the long-term evolution of fluvially based landforms.

ACKNOWLEDGEMENTS

This research was carried out while in receipt of a Natural Environment Research Council studentship at St Andrews University. The author thanks Dr Alan Werritty for helpful comments on an earlier draft of the paper and Mr Graeme Sandeman for redrawing the original diagrams.

REFERENCES

Acreman, M. C. (1983). The significance of the flood of September, 1981 on the Ardessie Burn, Wester Ross, *Scottish Geographical Magazine*, **99**, 150-60.

Anderson, M. G., and Calver, A. (1977). On the persistence of landscape features formed by a large flood, *Transaction of the Institute of British Geographers* (NS), **2**, 243-54.

Baird, P. D., and Lewis, W. V. (1957). The Cairngorm floods, 1956: summer solifluction

and distributary formation, *Scottish Geographical Magazine,* **73**, 91-100.
Bremner, A. (1922). The Dee floods of October, 1920, *The Deeside Field,* **1**, 30-2.
Brice, J. C. (1960). Index for description of channel braiding. *Bulletin of the Geological Society of America,* **85**, 581-6.
Carling, P. (1986), The Noon Hill flash floods; July 17th 1983. Hydrological and geomorphological aspects of a major formative event in an upland landscape, *Transactions of the Institute of British Geographers,* **11**, 105-18.
Carr, A. B. (1962). Cartographic record and historical accuracy, *Geography,* **47**, 135-44.
Church, M., and Ryder, J. M. (1972). Paraglacial sedimentation: a consideration of fluvial processes conditioned by glaciation, *Bulletin of the Geological Society of America,* **83**, 3059-72.
Costa, J. E. (1974). Stratigraphic, morphologic and pedologic evidence of large floods in humid environments, *Geology,* **2**, 301-3.
Harley, J. B. (1975). *Ordnance Survey Maps: a Descriptive Manual,* Ordnance Survey, Southampton.
Hickin, E. J. (1983). River channel changes: retrospect and prospect, in J. D. Collinson and J. Lewin, (eds), *Modern and Ancient Fluvial Systems,* Blackwell Scientific, London, pp. 61-83.
Hooke, J. M. (1977). The distribution and nature of changes in river channel patterns: the example of Devon, in K. J. Gregory, (ed.), *River Channel Changes,* Wiley, Chichester, pp. 265-280.
Hooke, J. M., and Kain, R. J. (1982). *Historical Changes in the Physical Environment: a Guide to Sources and Techniques,* Butterworths, London.
Jousley, P. C. (1973). Peatlands, in J. Tivy, (ed.), *The Organic Resources of Scotland,* Oliver and Boyd, Edinburgh, pp. 109-21.
Kellerhals, R., Church, M., and Bray, D. I. (1976). Classification and analysis of river processes, *Journal of the Hydrological Division of the American Association of Civil Engineers,* **102**, Proceedings Paper 12232, 813-29.
Lauder, T. D. (1830). *An Account of the Great Floods of August, 1829 in the Province of Moray and Adjoining Districts,* MacGillvray and Son, Elgin.
McEwen, L. J. (1986). *River Channel Planform Changes in Upland Scotland, with Specific Reference to Climatic Fluctuation and Land-use Changes over the Last 250 years,* University of St Andrews, PhD (unpublished).
McEwen, L. J. (1987a). Sources for establishing a historic flood chronology (pre-1940) within Scottish river catchments, *Scottish Geographical Magazine,* **103**, 121-40.
McEwen, L. J. (1987b). The use of long-term rainfall records for augmenting historic flood series: a case study on the upper Dee, Aberdeenshire, *Transactions of the Royal Society of Edinburgh: Earth Sciences,* **78**, 275-85.
McEwen, L. J., and Werritty, A. (1988). The hydrology and long-term geromorphic significance of a flash flood in the Cairngorm mountains, Scotland, *Catena,* **15**, 361-377.
Maizels, J. K. (1983). Channel changes, palaeohydrology and deglaciation: Evidence from some Lateglacial deposits of north-east Scotland, *Quaternary Studies in Poland,* **4**, 171-87.
Milne, J. A. (1982). *River Channel Change in the Harthorpe Valley, Northumberland, Since 1897,* University of Newcastle upon Tyne, Department of Geography Research Series 13.
Moir, I. (1973). *The Early Maps of Scotland to 1850 with a History of Scottish Maps: Vol. 1,* Royal Scottish Geographical Society, Edinburgh.
Natural Environment Research Council (1975). *Flood Studies Report,* London (5 volumes).

Patton, P. C., and Baker, V. B. (1977). Geomorphic response of Central Texas stream channels to catastrophic rainfall and runoff, in D. O. Doechring, (ed.), *Geomorphology in Arid Regions*, pp. 189–220, Binghampton, New York.

Pears, N. V. (1968). Man in the Cairngorms: a population-resource balance problem, *Scottish Geographical Magazine*, **84**, 45–55.

Roberts, C. H. (1919). Investigations of the Flows of the River Dee, Scotland, *Institution of Water Engineers* (Report).

Schick, A. P. (1974). Formation and obliteration of desert stream terraces—a conceptual analysis, *Zeitschrift fur Geomorphologie* Supplement **21**, 88–105.

Schumm, S. A. (1977). *The Fluvial System*, Wiley, New York.

Siegel, S. (1956). *Non-parametric statistics for the behavioural sciences*. McGraw-Hill, New York.

Sissons, J. B. (1976). *The Geomorphology of the British Isles: Scotland*, Methuen, London.

Sissons, J. B., and Walker, M. J. C. (1974). Lateglacial site in the central Grampian Highlands, *Nature*, **249**, 822–4.

Skelton, R. A. (1967). The military survey of Scotland, 1747–1755, *Scottish Geographical Magazine*, **83**, 5–16.

Steven, H. M., and Carlisle, A. (1959). *The native pinewoods of Scotland*, Oliver and Boyd, Edinburgh.

Tricart, J. (1961). Mechanismes normeaux et phénomènes catastrophiques dans l'évolution des versants du Basin de Guil, Hautes Alps, France, *Zeitschrift für Geomorphologie*, **5**, 277–301.

Werritty, A. (1984). Stream response to flash floods in upland Scotland, in D. E. Walling and T. P. Burt, (eds), *Catchment experiments in fluvial geomorphology* Geobooks, Norwich, pp. 537–60.

Werritty, A., and Ferguson, R. I. (1980). Pattern changes in a Scottish braided river over 1, 30 and 200 years, in R. A. Cullingford, D. A. Davidson and J. Lewin, (eds), *Timescales in Geomorphology* Wiley, Chichester, pp. 53–68.

Wolman, M. G., and Gerson, R. (1978). Relative scales of time and effectiveness of climate in watershed geomorphology, *Earth Surface Processes*, **3**, 189–208.

14 Sedimentology and Palaeohydrology of Holocene Flood Deposits in Front of a Jökulhlaup Glacier, South Iceland

JUDITH MAIZELS
Department of Geography, University of Aberdeen, Aberdeen

INTRODUCTION

This paper examines the sedimentology of a series of deposits associated with catastrophic floods caused by subglacial volcanic eruptions at Solheimajökull in southern Iceland. The aims of the study were to assess (1) the nature and dynamics of these flood flows; (2) the magnitude and frequency of these events; and (3) their geomorphological significance in modifying the proglacial sandur environment. This paper concentrates on the sedimentary characteristics of a distinctive series of lithofacies types and sedimentary sequences which are interpreted as having been formed during a number of Holocene glacier burst events.

PREVIOUS FLOOD RECORD IN SOLHEIMAJÖKULL STUDY AREA

Solheimajökull is an 8 km long outlet glacier which descends from the Myrdalsjökull ice cap in southern Iceland (Figure 14.1). The ice cap is underlain by the volcano Katla, which has erupted at least 17 times in recorded history (Thorarinsson, 1975; Einarsson *et al.*, 1980). The eruptions have produced extensive ash deposits, now recognised as distinct tephra layers in the stratigraphic record (Maizels and Dugmore, 1985; Dugmore, 1987), as well as a range of fluvial and debris flow deposits associated with the catastrophic floods, or 'jökulhlaups' extending across the proglacial sandur plains (e.g. Jonsson, 1982). The jökulhlaups have produced estimated discharges of up to $10^5 \, m^3 \, s^{-1}$ (e.g. Kotluhlaup 1918: Thorarinsson, 1957; Björnsson, 1975), while sediment loads of up to $7 \times 10^4 \, kg$ have been recorded during discharges of $c \, 2 \times 10^3 \, m^3 \, s^{-1}$ on Skeidarasandur, i.e. sediment concentration of $c. \, 35\%$ (Tomasson *et al.*, 1980). Most of this sediment comprises subglacially derived silts, sands and fine gravels

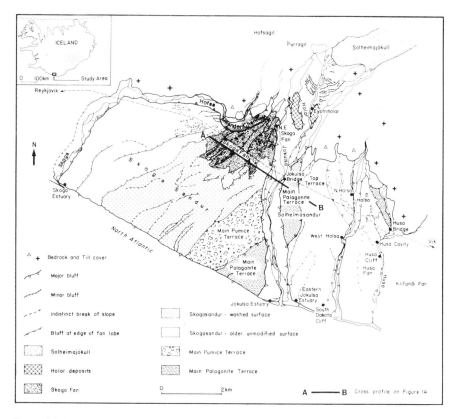

Figure 14.1 Location of study area and main morphological units of the Solheimajökull sandur area.

of black pumice, and boulders of yellow-brown palagonite tuffs and breccias scoured from deeply dissected flood routeway gorges. Flood sediments appear to comprise less than 2% clay (Tomasson, 1974; Tomasson et al., 1980), while containing large amounts of airfall tephra (Einarsson et al., 1980).

The extensive black pumice sand and gravel plains of Solheimasandur, Skogasandur and Myrdalsandur may therefore owe their existence largely to relatively infrequent, subglacially generated, high-magnitude flood events, rather than to long-term aggradation in the braided stream environments associated with the normal summer melting of these southern Iceland glaciers. This study aims to determine the nature and dynamics of flows associated with sediments within the sandur stratigraphic sequence, and hence to assess the relative importance of flood flows in creating the proglacial sandur plains.

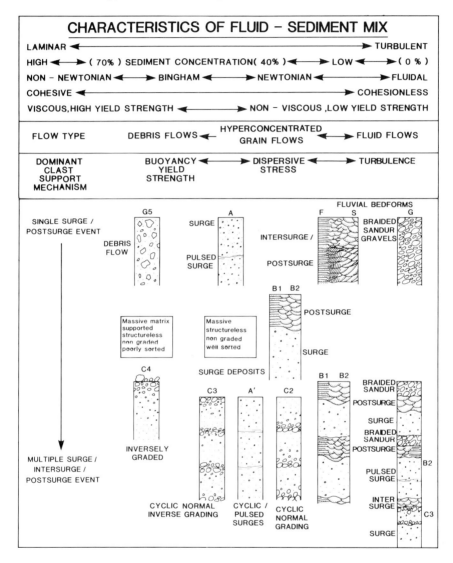

Figure 14.2 Simplified model linking flood flow conditions and flow characteristics to likely sedimentary structures of the resulting deposits. (See Figure 14.4 for explanation of symbols).

PREDICTED CHARACTERISTICS OF SOLHEIMASANDUR FLOOD DEPOSITS

Interpretation of the Solheimasandur sedimentary sequence has been based on prediction of the types of sediment that are likely to accumulate during these volcanically triggered flood flows compared with those forming during the relatively low-magnitude diurnal flows typical of the braided meltwater outlet streams. The high sediment concentration, low to moderate clay content, and low bulk densities ($1.6-1.8\,\mathrm{g\,cm^{-3}}$) of the pumice sands and fine gravels, are all likely to decrease the viscosity, dampen flow turbulence, and increase the yield strength of the deposit and the buoyancy of the constituent particles (e.g. Fisher, 1971; Hampton, 1972, 1979; Wasson, 1977; Pierson, 1981; Postma, 1986).

The likely relationships between sediment concentration, flow type and sedimentary structures for Solheimajökull flood deposits have been portrayed in a tentative model (Figure 14.2). The model indicates that sediments with high sediment-water ratios are likely to result in non-Newtonian (pseudoplastic) behaviour, acting as a cohesive debris flow, and producing massive, matrix-supported, poorly sorted, lobate deposits (Type G5 in Figure 14.2) (e.g. Takahashi, 1980; Johnson, 1970; Johnson and Rahn, 1970; Nemec and Steel, 1984). Where flows form highly concentrated cohesionless (grain) flows, with >40% sediment concentration (Beverage and Culbertson, 1964; Lowe, 1976; Costa, 1984; Pierson and Costa, 1987), massive, structureless sands and gravels may be deposited following rapid 'freezing' of the material during fluid expulsion (Type A in Figure 2). Dispersive pressure associated with Bingham behaviour (Bagnold, 1954; Enos, 1977; Pierson, 1981) can also cause the coarser grains to migrate to zones of lower shear, promoting the development of inversely graded deposits (e.g. Fisher, 1971; Larsen and Steel, 1978; Nemec and Steel, 1984), with larger clasts occurring as 'rafted' surface and lobe-edge boulders (e.g. Hampton, 1979; Pierson, 1981) (Type C4 in Figure 14.2). These sediment types contrast markedly with those characteristic of turbulent, fluid flows associated with low sediment concentrations, which are likely to produce fluvial bedforms (ripples, dunes, ribs) and imbricated bedload clasts (top right-hand zone of Figure 14.2). Where turbulent flows contain high sediment concentrations, shearing produces crude sub-parallel laminations, imbricated slurries, basal erosional scours and flutes, and possible inverse grading (e.g. Sanders, 1965; Enos, 1977).

THE SOLHEIMASANDUR DEPOSITS

Eleven main stratigraphic units have been identified in the Solheimasandur deposits (Figure 14.3). These include a wide range of lithofacies (Table 14.1) which were found to occur in three main vertical associations or sequence types (Figure 14.4).

Lithofacies

Fines. Fine-grained units are relatively uncommon in the stratigraphic sequence, and are confined to two types of stratigraphic position. First, units of

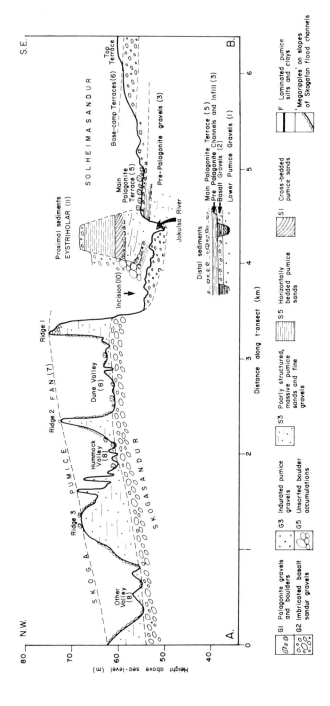

Figure 14.3 Schematic diagram illustrating main stratigraphic relationships of pre-Little Ice Age sandur deposits, Solheimajökull.

Table 14.1 Sedimentary characteristics of dominant lithofacies in Solheimajökull sandur deposits.

	Description	D50 (mm)	D95 (mm)	Sedimentary characteristic Sorting index (mm)	Max % silt + clay	Max unit thickness (m)	Solid density (g cm^{-3})
Fines							
F2	Load-structured silts and clays	0.03	0.08	0.60	90.0	2.80	1.60
F1	Laminated silts and fine sands	0.06	0.16	0.58	58.0	0.26	1.60
Sands							
S8	Deformed and faulted, bedded, pumice sands and fine gravels	1.40	9.1	0.17	3.8	0.85	—
S7	Sands with climbing ripple sequences	1.30	4.9	0.34	0.4	0.50	—
S6	Cross-bedded black pumice sands	1.36	6.01	0.29	1.4	0.45	—
S5	Horizontally laminated, coarse, black silts and sands	0.48	4.48	0.21	11.3	3.6+	—
S4	Plane-bedded sands and fine gravels	0.68	4.49	0.29	3.2	3.5+	—
S3	Poorly structured, massive, black pumice sands and gravels	1.30–2.77	8.0	0.21–0.34	3.9	4.6+	1.64
S2	Fining-upwards, cross-bedded, pumice sands and gravels	1.82	7.3	0.34	3.8	0.6	—
S1	Trough cross-bedded, pumice sands and gravels	1.74	7.2	0.26	5.3	0.45	—
Gravels							
G9	Clast-supported, structureless gravels	2.35	8.1	0.27	4.5	4.0+	—
G8	Pebble stringer	1.95–28.84	15.9–43.4	0.23–0.34	0.7–3.5	0.4	—
G7	Matrix-supported, unsorted, angular, heterogeneous mix in sandy matrix, with flow structures and cavities	1.40	9.1	0.17	3.8	0.9	—
G6	Clast-supported, crudely bedded, heterogeneous, rounded, sandur gravels	4.68	40.4	0.22	0.01	1.9	~2.50
G5	Matrix-supported, unsorted, angular, heterogeneous mix in clay-rich matrix	0.93	12.8	0.16	68.0	1.9	—
G4	Cross-bedded, basalt-rich gravels	8.11	35.5	0.22	1.3	3.0+	—
G3	Clast-supported, indurated, pumice gravels in white/yellow tephra matrix	1.48–3.51	6.4–14.7	0.28–0.36	2.6	2.5+	1.50
G2	Clast-supported, imbricated, basalt-rich gravels and pebbles	15.31	42.9	0.21	1.3	1.9	2.50
G1	Clast-supported, palagonite cobbles and boulders in black pumice matrix	7.89	(<4m)	0.13	3.5	4+	1.85

black, laminated silts and fine sands (F1, Figure 14.4) up to 20 cm thick, occur at the base of scoured channels, and second, load-structured silts and clays (F2) occur at the contact with overlying thick silt and sand beds.

Sands. Eight distinct sand lithofacies have been identified, dominating the Solheimasandur depositional record. These lithofacies are all composed of black pumice sands, with less than 3% clay present. The thinnest units (<0.5 m) are those that exhibit cross-bedding, ranging from planar and trough cross-bedded pumice sands and gravels (S6, S1), with fining-upwards laminae (S2), to pumice sands with climbing ripple sequences (S7). The greatest thicknesses of units, exceeding 4.5 m in some sections, are exhibited by massive, homogeneous, poorly structured, well-sorted, fine black pumice sands and fine gravels (S3), and by finely horizontally laminated, coarse pumice silts and sands (S5), and plane-bedded pumice sands and fine gravels (S6) (Table 14.1). The homogeneous sands and fine gravels also exhibit indistinct sub-horizontal bedding or shear planes (?) at vertical intervals of 0.5–1 m, and occasional palagonite pebble stringers of single clast thicknesses (G8, below).

Gravels. The gravel lithofacies are dominated by thick units (3.5 m+) of pumice gravels, which become increasingly bleached and indurated with age (G3), and are associated with palagonite clasts, either forming the matrix of palagonite cobble gravels, or being directly overlain by palagonite boulder beds (G1). The pumice gravels generally form thick, clast-supported, structureless, non-imbricated deposits (G9). The G9 gravels contrast markedly with the thin beds (<1 m) of crudely bedded, heterogeneous or basalt-rich, rounded sandur gravels (G6, G4, G2) which separate the main pumice gravel sequences. Matrix-supported gravels occur only locally, comprising unsorted, angular, heterogeneous clasts in a clay-rich ($<68\%$ silt + clay; G5) or sandy matrix (G7), the latter exhibiting deformed bedding, flow structures and cavities (Table 14.1).

LITHOFACIES SEQUENCES

The Solheimasandur pumice lithofacies exhibit three main sequence types (Figure 14.4).

Sequence A comprises the massive, homogeneous, well-sorted, fine black pumice sands and gravels of lithofacies S3, G3 and G9. These lithofacies dominate large areas of the sandur deposits, reaching thicknesses of over 4.6 m at many sites.

Sequence B1 is composed of basal sediment units similar to those of Sequence A, but these gravels are capped by an erosional surface and overlain by

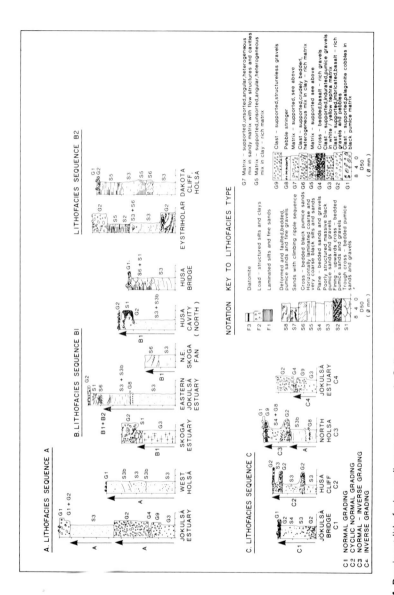

Figure 14.4 Dominant lithofacies sediment sequences in Sólheimajökull flood deposits.
(a) *Sequence A*: Massive, homogeneous, well-sorted, fine black pumice sands and gravels.
(b) *Sequence B*: Massive, homogeneous, well-sorted, fine black pumice sands and gravels, capped by erosion surface and thin cross-bedded (B1) or plane-bedded (B2), black pumice sands and gravels.
(c) *Sequence C*: Graded sediments:
  C1 Normal grading
  C2 Cyclic–normal grading
  C3 Normal–inverse grading
  C4 Inverse grading

a series of thin cross-bedded, black pumice sands and fine gravels of lithofacies S1, S2, S6 and G4. The basal gravels of these upper cross-bedded units comprise a coarser clast layer, but the infill gravels remain poorly graded and massive in structure. These gravels are commonly mantled by large (<2 m) palagonite boulders. In addition, the uppermost cross-bedded units are overlain along the routes of former flood channels by transverse surface dunes. The dunes have wavelengths averaging between 6.6 and 10.5 m, and comprise fine gravel cores but are mantled with a lag of basalt clasts up to 25 mm in diameter. *Sequence B2* is represented by Sequence A sediments overlain by thick (<3.5 m), horizontally bedded, black, pumice sands and fine gravel units of lithofacies S4 and S5 (Figure 14.4). These units comprise thin (averaging 3 cm), alternating coarse and fine laminae (with median sizes of *c.* 1.3 and 0.06 mm, respectively), low percentages of silt and clay (<0.5 and <4.6% respectively), and isolated clasts up to 4 cm in diameter (Table 14.1).

Sequence C is represented by a range of graded sediments, including normal (C1), cyclic–normal (C2), normal–inverse (C3), and inverse grading (C4) (Figure 14.4). These units commonly exceed 3 m in thickness, contain <5.6% clay, and include both matrix- and clast-supported gravels (Table 14.1). Sequence C1 comprises fining-upwards sands and gravels, with non-pumice basal clasts, and a mean thickness of *c.* 1 m. Sequence C2 comprises units up to 2 m thick of cyclic–normal graded sediment, in which each cyclic unit exhibits a basal, clast-supported layer grading vertically into a homogeneous and poorly structured fine gravel unit. At least three such cyclic units have been identified within the C2 sequences. Sequence C3 consists of normal–inverse graded units up to 1.5 m thick, in which a coarse (<25 cm diameter) basal layer of clast-supported, poorly sorted, and weakly imbricated gravels grades upwards, first into a crudely bedded, homogeneous, fine gravel unit containing scattered clasts, and second, into an uppermost clast-supported pebble bed. This upper coarse zone is finer grained and contains a higher matrix content than occurs in the basal clast layer. Sequence C4 comprises inversely graded pumice gravels, averaging 1 m in thickness, in which well-sorted, fine-grained, pumice gravels grade upwards into more massive, heterogeneous, subrounded, pebble gravels.

PALAEOFLOW INTERPRETATION

The massive, structured, homogeneous or inversely graded nature of the dominant lithofacies types present in the Solheimasandur deposits suggests that they were deposited from hyperconcentrated fluid–sediment mixtures (cf. Figure 14.2). The indistinct 'pseudobedding' and cyclic grading (e.g. C2, C3, C4) are likely to have resulted from flows occurring as a series of pulses (e.g. cf. Pierson 1980), while dispersive stresses allowed the development of inverse grading and rafting of boulders (e.g. C4). The dominant lithofacies sequence (Sequence B), in which thick, massive, poorly structured, pumice sands and gravels forming lobate features are overlain by thin units of cross-bedded, horizontally bedded, graded or poorly graded sands and gravels, suggests that significant changes in flow conditions occurred during the period of deposition. The lower massive deposits are interpreted as representing an initial flood surge associated with the jökulhlaup peak, which arrived as a series of pulsed flows. The flood waters transported high concentrations of pumice sands and gravels derived directly from the volcanic eruption, together with ice and palagonite boulders, to form a hyperconcentrated and highly competent fluid. Peak flows were followed by flow deceleration, sediment depletion (with cessation of the eruption), and dewatering as the fan lobe ($<$11 m high) came to a halt. A more fluid series of flows developed, associated with erosion, to form deep ($<$12 m), wide ($<$500 m) channels and fields of streamlined residual hummocks, and with the formation of low flow regime, fluvial dunes (cf. Smith, 1987).

MAGNITUDE AND FREQUENCY OF FLOOD EVENTS

The identification of the main flow characteristics likely to have been associated with each deposit (from Figure 14.2) provides a more valid basis for the selection of the most appropriate models for estimating former flow magnitudes (e.g. see Costa 1984). However, because of uncertainties regarding calculations of former sediment concentration, bulk density, yield strength and viscosity of the flood flows (but see Maizels, 1989), no model can yet provide very accurate estimates of palaeoflow magnitudes. The present calculations of flows have been based on a range of published models (Bagnold, 1954; Johnson, 1970; Enos, 1977; Costa, 1984) requiring inputs of a large range of particle and fluid characteristics. Error sources are attached to the determination of each of the listed parameters, such that the final flow estimates are likely to be correct only to the nearest order of magnitude (see Maizels, 1989). The validity of the results can only be assessed at this stage by comparison with published values of velocity, Froude number, and discharge for a range of monitored fluid and hyperconcentrated flood flows (e.g. Pierson, 1980, 1985; Costa, 1984).

Estimates of peak flows associated with the pumice gravel deposits suggest that velocities ranged between about 3.9 and 13 m s^{-1}, with maxima of $c.$ 17 m s^{-1}; peak discharges ranged from 3.3×10^5 m^3 s^{-1} during flood surges to only 9.1×10^2 m^3 s^{-1} during the more fluid phases. These flow estimates fall well within the range of velocities for hyperconcentrated flows recorded by a number of other workers (see summaries, Costa, 1984), but tend toward the lower part of the range (cf. Pierson, 1980, 1985), reflecting the relatively low gradients (0.035 to 0.003) of the sandur plains.

The dating of the older deposits remains tentative, but the application of tephrochronology, ^{14}C dating, and relative weathering indices (Maizels and Dugmore, 1985; Dugmore, 1987), suggest that at least eight major jökulhlaups have occurred on Solheimasandur over the period between c. 4500 years BP and AD 1357.

GEOMORPHIC SIGNIFICANCE OF JOKULHLAUP EVENTS

The series of jökulhlaup events recorded in the sedimentary and morphological record at Solheimasandur appears to have acted as a dominant control on long-term sandur development, both in terms of depositional and erosional processes. In depositional terms, particularly, the extensive lobes, fans and gravel sheets of this part of the south Iceland coastal plain are composed of great thicknesses ($>$ 100 m, e.g. Boulton et al., 1982) of dominantly jokulhlaup sediments. Indeed, the Solheimasandur sedimentary record suggests that over 85% of the total thickness of proglacial sandur deposits is composed of jökulhlaup flood sediments. These deposits have locally been dissected into a series of river terraces and major flood channels, leaving residual streamlined hummocks scattered with boulder lags and relic dunes. Hence, the jökulhlaup-modified landscape is one of substantial relative relief, with local dissection into the sandur deposits of up to 30 m.

CONCLUSIONS

Although major flood events have been relatively rare on Solheimasandur, with only eight recorded in the sedimentary record over a 4000 year period, their effects still dominate the proglacial landscape. These large flows, probably exceeding 10^4–10^5 m^{3}s^{-1}, have produced repeated sequences of distinctive, massive, homogeneous, black pumice gravels, characteristic of hyperconcentrated flows, overlain by erosional surfaces, boulder lags, and fluvial bedforms. This common sequence is interpreted as a surge–postsurge flow sequence developed during a complex, volcanically generated jökulhlaup event. Major erosional events occurred during the later flood stages, creating large flood channels and possibly initiating terrace incision.

Many problems and uncertainties remain in the interpretation of the wide range of sedimentary structures present in the flood deposits, both in terms of former instantaneous flow conditions, and changes of flow during transport and deposition. The overall approach adopted here may provide a useful framework for the more detailed analysis of flood deposits which have accumulated under a wide range of sediment–fluid flow conditions.

REFERENCES

Bagnold, R. A. (1954). Experiments on a gravity-free dispersion of large solid spheres in a Newtonian fluid under shear, *Proc. Roy. Soc. Lond.*, Series A, **225**, 49–63.

Beverage, J. P., and Culbertson, J. K. (1964). Hyperconcentrations of suspended sediment, *J. Hydr. Div., Amer. Soc. Civil Engr.*, **HY6**, 117–26.

Bjornsson, H. (1975). Subglacial water reservoirs, jokulhlaups and volcanic eruptions, *Jokull*, **25**, 1–14.

Boulton, G. S., Harris, P. W. V., and Jarvis, J. (1982). Stratigraphy and structure of a coastal sediment wedge of glacial origin inferred from sparker measurements in glacial Lake Jokulsarlon in southeastern Iceland, *Jokull*, **82**, 37–47.

Costa, J. E. (1984). Physical geomorphology of debris flows, in J. E. Costa and P. J. Fleisher, (eds), *Developments and Applications of Geomorphology*, Springer-Verlag, pp. 268–317.

Dugmore, A. J. (1987). *Holocene Glacier Fluctuations Around Eyjafjallajokull, Southern Iceland*, Unpublished Ph.D. thesis, University of Aberdeen.

Einarsson, E. H., Larsen, G., and Thorarinsson, S. (1980). The Solheimar tephra layer and the Katla eruption of ~1357, *Acta Nat. Isl.*, **28**, 24 pp.

Enos, P. (1977). Flow regimes in debris flows, *Sedimentol.*, **24**, 133–42.

Fisher, R. V. (1971) Features of coarse-grained, high concentration fluids and their deposits, *J. Sed. Pet.*, **41**, 916–27.

Hampton, M. A. (1972). The role of subaqueous debris flow in generating turbidity currents, *J. Sed. Pet.*, **42**, 775–93.

Hampton, M. A. (1979). Buoyancy in debris flows, *J. Sed. Pet.*, **49**, 753–8.

Johnson, A. M. (1970). *Physical Processes in Geology*, Freeman and Cooper, San Francisco, 577 pp.

Johnson, A. M., and Rahn, P. H. (1970). Mobilization of debris flows, *Zeit. Geomorph.*, Suppl. Bd. **9**, 168–86.

Jonsson, J. (1982). Notes on the Katla volcanological debris flows, *Jokull*, **32**, 61–8.

Larsen, V., and Steel, R. J. (1978). The sedimentary history of a debris flow-dominated alluvial fan: a study of textural inversion, *Sedimentol.*, **25**, 37–59.

Lowe, D. R. (1976). Grain flow and grain flow deposits, *J. Sed. Pet.*, **46**, 188–99.

Maizels, J. K. (1987). Large-scale flood deposits associated with the formation of coarse-grained braided terrace sequences, in Recent Developments in Fluvial Sedimentology, F. G. Ethridge, R. M. Flores, and M. D. Harvey, (eds), *Soc. Econ. Pal. Min.*, Spec. Publn. **39**, 135–48.

Maizels, J. K. (1989). Sedimentology, paleoflow dynamics and flood history of jokulhlaup deposits: paleohydrology of Holocene sediment sequences in southern Iceland sandur deposits, *J. Sed. Pet.*, **59**.

Maizels, J. K., and Dugmore, A. J. (1985). Lichenometric dating and tephrochronology of sandur deposits, Solheimajokull area, south Iceland, *Jokull*, **35**, 69–77.

Nemec, W., and Steel, R. J. (1984). Alluvial and coastal conglomerates: their significant features and some comments on gravelly mass-flow deposits, in E. H. Koster and R. J. Steel, (eds), *Sedimentology of Gravels and Conglomerates*, Canad. Soc. Petrol. Geol. Memoir 10, pp. 1–31.

Pierson, T. C. (1980). Erosion and deposition by debris flows at Mount Thomas, north Canterbury, New Zealand, *Earth Surf. Proc.*, **5**, 227–47.

Pierson, T. C. (1981). Dominant particle support mechanisms in debris flows at Mount Thomas, New Zealand, and implications for flow mobility, *Sedimentol.*, **28**, 49–60.

Pierson, T. C. (1985). Initiation and flow behaviour of the 1980 Pine Creek and Muddy River lahars, Mount St. Helens, Washington, *Bull. Geol. Soc. Amer.*, **96**, 1056–96.

Pierson, T. C. and Costa, J. E. (1987). A rheologic classification of subaerial sediment-water flows, *Geol. Soc. Amer., Reviews in Engr. Geol.,* **VII**, 1–12.
Postma, G. (1986). Classification for sediment gravity-flow deposits based on flow conditions during sedimentation. *Geology,* **14**, 291–4.
Sanders, J. E. (1965). Primary sedimentary structures formed by turbidity currents and related re-sedimentation mechanisms, *Soc. Econ. Pal. Min.,* Spec. Publn. **12**, 192–19.
Smith, G. A. (1987). The influence of explosive volcanism on fluvial sedimentation: the Deschutes Formation (Neogene) in Central Oregon, *J. Sed. Pet.,* **57**, 613–29.
Takahashi, T. (1980). Debris flow on prismatic open channel, *Proc. Amer. Soc. Civil. Engr., Jl. Hydr. Div.,* **106** (HY3), 381–96.
Thorarinsson, S. (1957). The jokulhlaup from the Katla area in 1955 compared with other jokulhlaups in Iceland, *Jokull,* **7**, 21–5.
Thorarinsson, S. (1975). Katla og annall Kotlugosa, *Arbok Ferdafelags Islands,* **1975**, 124–49.
Tomasson, H. (1974). Grimsvatnahlaup 1972. Mechanism and sediment discharge, *Jokull,* **24**, 27–39.
Tomasson, H., Palsson, S., and Ingolfsson, P. (1980). Comparison of sediment load transport in the Skeidara jokulhlaups in 1972 and 1976, *Jokull,* **30**, 21–33.
Wasson, R. J. (1977). Last-glacial alluvial fan sedimentation in the Derwent Valley, Tasmania, *Sedimentol.,* **24**, 781–99.

Floods: Hydrological, Sedimentological and Geomorphological Implications
Edited by K. Beven and P. Carling
© 1989 John Wiley & Sons Ltd

15 Flood Deposits Present within the Severn Main Terrace

MARTIN DAWSON
Department of Geography, University College of Wales, Aberystwyth
(Present address: BP International Ltd, Britannic House, Moor Lane, London)

INTRODUCTION

Although the sedimentary sequences of coarse grained, low sinuosity Quaternary fluvial deposits have been well documented, (Costello and Walker, 1972; Eynon and Walker, 1974; Frazer, 1982; Bryant, 1983; Thomas *et al.*, 1985; Billi *et al.*, 1987) it has rarely been possible to identify distinct, event related, depositional bodies within such sequences. The purpose of this paper is to describe and interpret such a sedimentary horizon, thought to be of flood origin, present within a Devensian, paraglacial, terrace deposit and to provide some general estimates of the flood magnitude.

The Severn Main Terrace is the best preserved of a suite of four terraces and up to three higher level terrace remnants bordering the lower River Severn (Wills, 1938; Hey, 1958; Beckinsale and Richardson, 1964). It is traceable as a geomorphic feature from Apley park near Bridgnorth, where it lies 30 m above the present floodplain, to Gloucester, where the deposits extend beneath the modern alluvium (Figure 1). The terrace surface is underlain by extensive sand and gravel deposits. These are locally traceable for up to 1 km across the valley and may have a thickness of up to 10 m. This coarse clastic aggradation developed as a response to a high sediment yield: discharge ratios during the Devensian glacial maximum (Dawson and Gardiner, 1987) and is dated at between 18 000 and 25 000 B.P. through a correlation with glacigenic diamict deposits found interstratified with the terrace deposits at Eardington (Shotton, 1977; Shotton and Coope, 1983; Dawson, 1985).

DEPOSITIONAL SETTING

The terrace aggradation has been interpreted as having been deposited by a low sinuosity gravel bed river which, at least locally, had a braided planform (Wills, 1938; Dawson, 1985; Dawson and Bryant, 1987). The most extensive terrace

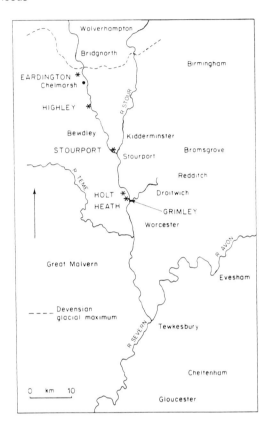

Figure 15.1 Location of the major exposures in the Severn Main Terrace.

remnant occurs at Holt Heath (Figure 15.1), where the sediment aggradation has a maximum thickness of 7 m and extends across the valley for up to 0.75 km. Here the terrace deposit has a distinctive architecture with discrete gravel and sand bodies being defined by major bounding surfaces (Figure 15.2) (Dawson and Bryant, 1987). This structure has been interpreted as having been produced through the repeated migration and avulsion of a gravelly braided channel zone across the valley floor under conditions of relatively high sediment accumulation (Dawson, 1986).

FLOOD SEQUENCE

One gravel body at Holt Heath, although interstratified with adjacent sand units like the other 'channel zone' bodies, shows sedimentary characteristics atypical

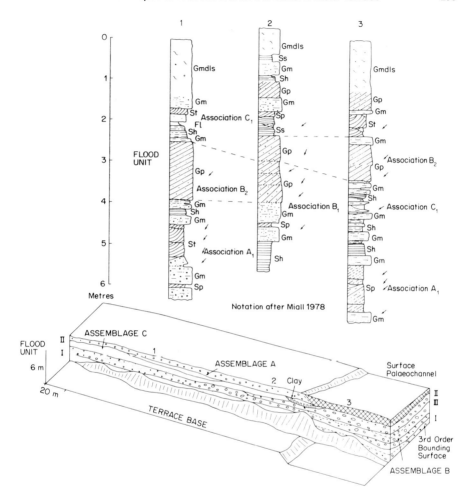

Figure 15.2 Sedimentary characteristics and architectural structure of the terrace sequence at Holt Heath.

of the sequence as a whole (Figure 15.3). It is traceable for 380 m, across much of the available terrace exposure, and rests above a distinct planar, erosional, surface. The discordant nature of the unit relative to underlying lithofacies is particularly evident where the surface truncates synsedimentary folding present in sand units beneath it (Dawson, 1986).

Similar sedimentary units are apparent within terrace deposits at Highley and Stourport (Figure 15.1). However, stratigraphic correlation between fragmentary terrace deposits, at unit level, is inadvisable given the complex controls on

Figure 15.3 Flood unit at Holt Heath showing surficial clay drape, cross - stratified gravel lithofacies, and basal erosional horizon.

terrace formation (Dawson and Gardiner, 1987) and it is uncertain whether the separate units may be regarded as the product of a single flood.

The main body of the unit is composed of gravel beds showing both crude planar and cross-stratification with two main lithofacies types present.

(1) *Crudely planar stratified beds up to 1.2 m thick.* Typically these beds are poorly stratified, framework supported and matrix rich. However, up dip of the cross-stratified lithofacies such beds show planar stratification comprising superimposed layers of openwork, imbricated, matrix rich, and disorganised matrix rich gravel 0.10–0.25 m thick. The contacts between such layers are rarely sharp and may have a welded (Steel and Thompson, 1983) appearance.

(2) *Units of low angle (15 to 20°), radially dipping cross-strata.* These beds are laterally traceable for up to 50 m as single units and have thicknesses exceeding 0.85 m. Usually the beds occur as isolated units, but, locally, they were observed to be multiply stacked. Preservation of the original geometry of the units is good: the cross-strata rest either conformably on the basal erosional surface, or on the underlying cross-strata, and the upper surfaces have a convex-up form, dipping in directions parallel and normal to the internal foreset orientation.

Figure 15.4 Eroded channel form at Holt Heath containing tabular cross-bedded gravel fill.

Internally, the units are composed of multiple bundles of irregularly dipping cross-strata, separated by low-angle dipping convex-up reactivation surfaces. Adjacent bundles are laterally discordant and show minor differences in dip angles and direction (maximum variation ± 15°). Individual foresets show normal grading of the following form:

Coarse sand
———Gradational———
Openwork fine gravel
———Welded———
Openwork medium gravel
———Sharp———
Disorganised matrix rich gravel
———Sharp———

This sequence is not present at all locations, the sand member often being truncated. The matrix of the disorganised matrix rich gravel often has a high clay content whilst sandy clay drapes are present on the upper surfaces of individual foresets, and, more commonly on the reactivation surfaces separating adjacent foreset bundles. The gravel components often show reverse tangential grading fining from the unit base towards the topsets. Clasts resting on the foresets lie with their A–B planes concordant to the dip of the foresets and there is no evidence of counter-dipping cross-strata indicating flow transverse to the orientation of the bedding (cf. Bluck, 1980).

Towards the valley sides the unit thins and there is a transition from gravel lithofacies to, predominantly, cross-stratified sandy lithofacies showing a similar (200–230°) palaeocurrent direction to the gravel lithofacies. The cross-stratification is bounded by erosive trough-like set bases 1–3 m wide, individual sets reaching up to 0.65 m in height. Migration of some of these bedforms must have been relatively slow as the foreset-base contacts are angular (Allen, 1968; 1982) and must have occurred under silt laden conditions as a silt drape forms the top member of individual cross-laminae.

Incised into the extensive gravel sheets are erosional channel forms 5–30 m wide. Two types of channel fill were observed.

(1) Coarse gravel fills, partially composed of tabular cross-stratified units oriented normal to the palaeocurrent direction of the main unit. These extend from the channel sides in wedge-like form and show both normally graded foresets and distinct fining up within the unit as a whole (Figure 15.4).

(2) Sandy channel fills composed of coarse, crudely laminated, pebbly sands and isolated, interstratified, tabular sets of cross-stratified sands. The cross-stratification, like the gravel units, tends to have a wedge-like form and is oriented normal to the channel axis.

The complete sequence is overlain by a clay drape 0.10–0.30 m thick. This is laterally extensive and is traceable along much of the length of the section. The drape shows no internal lamination, although where it overlies the erosive channel forms the drape thickens and may become interstratified with the sandy channel fills.

INTERPRETATION

The main gravel unit was the product of gravel sedimentation in, at least, two bed configurations. The crudely planar stratified gravel seems to have aggraded in sheet form, variations in the internal grading being produced by minor flow fluctuations (Smith, 1974; Rust, 1984) or bedload pulsation (Reid and Frostick, 1987). However, the cross-bedded lithofacies indicates the existence of sizeable channel floor bedforms. The convex-up nature of the major reactivation and bounding surfaces within the units suggest that, in long profile, the bedforms may have had a humpback form (Allen, 1983). Additionally, the divergence of the foreset dip direction from the valley axis orientation and the cross-stratified internal structure seems to indicate a 'lateral-diagonal' bar morphology (Billi *et al.*, 1987).

Various explanations have been proposed to account for textural differentiation in the foresets of gravel bars. Here, the evidence supports the contention of Carling and Glaister (1987) that the vertical grading within individual foresets is the result of quasi-simultaneous deposition of contrasting sub-facies on different areas of the bar front under non-uniform steady flow. Such variation, they argued, is the consequence of spatial and temporal differences in the local flow structure associated with progressive bar-front progradation. Additionally, the reverse tangential grading seen within the gravel members of the foresets may be explained by the particle overpassing mechanism (Allen, 1983) observed within their experimental situation.

The presence of multiple reactivation surfaces and the irregular nature of the foresets indicates that the textural differentiation may additionally be the result of unsteady flow conditions (Smith, 1974; Steel and Thompson, 1983; Rust 1984). Alternatively, such variation may be the consequence of sorting on the bar surface (Baker, 1973; Steel and Thompson, 1983; Allen, 1982; Bluck, 1984) or the result of the migration of minor bedforms over the bar surface and their subsequent collapse at the bar margins (Rust, 1984). However, as suggested by Carling and Glaister (1987) the textural differentiation in the foresets, in this case, cannot be related to avalanching at the bar margin (Steel and Thompson, 1983; Rust, 1984; Billi *et al.,* 1987) as the foreset dip angles are well below the angle of repose.

The lateral transition from gravel lithofacies to sandy cross-stratified lithofacies above the same erosional horizon indicates a change in bed conditions across the area of flow. Such a transition may be regarded as a consequence of an expansion of relatively low-velocity, silt-laden, flow onto an area of previously deposited, and exposed, sandy lithofacies, which became eroded and reworked.

The incised channel forms, infilled with both sand and gravel lithofacies, post-date the deposition of the laterally extensive gravel units. Their form is the product of a local concentration of flow, from the earlier, deeper and more extensive sheet-like form. However, some flow must have continued over adjacent areas in order to produce the channel fills which resulted from sediment movement both along the channel axis, and from the channel banks. The gravel fills clearly post-date the event maximum and seem to have been produced at an early stage of a general waning of flow depths and velocities, whilst the sandy fills seem to be an even later phenomenon. The presence of a clay drape capping the depositional sequence indicates a general decline in flow velocities and deposition from suspension.

The sequence of lithofacies as a whole seems to represent a major flood deposit. This is indicated by:

(1) the abrupt and unconformable transition from underlying units, indicating extensive erosion and scour;

CONCLUSIONS

Detailed sedimentological analysis of a paraglacial terrace deposit at Holt Heath enabled the identification of a sedimentary horizon atypical of the sequence as a whole. Examination of this horizon revealed the preservation of channel zone flood bedforms, a reworked overbank area, and waning flow features such as channel fills and an extensive clay drape. On this evidence the unit was interpreted as a flood deposit. Palaeodischarge calculations, although associated with a large range of possible error, indicated that the discharge at Holt Heath was considerable, possibly 2.5 times greater than the present day flood maximum. Episodic high magnitude flow events in modern paraglacial environments have been widely described (Arnborg, 1955; Church, 1978; Bjornson, 1975; Jonnsson, 1982), associated with sudden discharge fluctuations at the ice margin during deglaciation (jokulhlaups), and it is thought the sedimentary unit described here is the local consequence of such an event occurring during the initial stages of the deglaciation of the Devensian ice sheet.

ACKNOWLEDGEMENTS

The field data for this study was collected whilst the author held a National Environment Research Council research studentship at the University of Leicester. The study benefited greatly from discussions with Dr Ian Bryant, whilst the interpretation of the sediments as a flood unit is the development of an idea suggested by Professor Denys Brunsden at the 1984 BGRG Spring Field Meeting.

REFERENCES

Allen J. R. L. 1968 *Current Ripples*, Elsevier, Amsterdam, 433 pp.
Allen, J. R. L. (1982). *Sedimentary Structures*, Vol. 1, Developments in Sedimentology No. 30A, Elsevier, Amsterdam, 593 pp.
Allen J. R. L. (1983). Gravel overpassing on humpbacked bars supplied with mixed sediment: Examples from the Lower Old Red Sandstone, Southern Britain, *Sedimentology*, **30**, 285-94.
Arnborg, L. (1955). Hydrology of the glacial river Austurfljot, *Geografiska Annaler*, **37**, 185-201.
Ashmore, P. E. (1982). Laboratory modelling of gravel braided stream morphology, *Earth Surface Processes and Landforms*, **7**, 201-25.
Baker, V. R. (1973). *Paleohydrology and Sedimentology of Lake Missoula Flooding in Eastern Washington*, Geological Society of America: Special Paper No. 114, 70pp.
Beckinsale, R. P., and Richardson, L. (1964). Recent findings on the physical development of the Lower Severn Valley, *Geographical Journal*, **130**, 87-105.
Billi, P., Magi, M., and Sagri, M. (1987). Coarse-grained, low sinuosity river deposits: Examples from the Plio-Pleistocene: Valdarno Basin, in F. G. Ethridge, R. M. Flores,

and M. D. Harvey. (eds) *Recent Developments in Fluvial Sedimentology*, Society of Economic Mineralogists and Paleontologists, Special Publication 39, 197–204.

Bjornsson, H. (1975). Explanation of jokulhlaups from Grimsvoth, Vanojokull, Iceland, *Jokull*, **24**, 1–24.

Bluck, B. J. (1979). Structure of coarse grained braided stream alluvium, *Transactions of the Royal Society of Edinburgh*, **70**, 181–221.

Bluck, B. J. (1980). Structure, generation and preservation of upward fining braided stream cycles in the Old Red Sandstone of Scotland, *Transactions of the Royal Society of Edinburgh*, **71**, 29–46.

Bluck, B. J. (1984). The texture of gravel bars in braided streams, in R. E. Hey, J. C. Bathurst and C. R. Thorne, (eds), *Gravel Bed Rivers: Fluvial Processes, Engineering and Management*, Wiley, New York.

Bryant, I. D. (1983). Facies sequences associated with some braided river deposits of late-Pleistocene age from southern Britain, in J. D. Collinson and J. Lewin, (eds), *Modern and Ancient Fluvial Systems: Sedimentology and Processes*, International Association of Sedimentologists, Special Publication 6, 267–275.

Carling, P. A., and Glaister, M. S. (1987). Rapid deposition of sand and gravel mixtures downstream of a negative step: The role of matrix infilling and particle-overpassing in the process of bar-front accretion, *Journal of the Geological Society, London*, **44**, 543–51.

Cheetham, G. H. (1976). Palaeohydrological investigations of river terrace gravels, in D. A. Davidson and M. Shackley, (eds), *Geoarchaeology: Earth Sciences and the Past*, Duckworths, London, pp. 335–44.

Cheetham, G. H. (1980). Late Quaternary palaeohydrology: The Kennet Valley case study, in D. K. C. Jones, (ed.), *The Shaping of Southern England* Institute of British Geographers Special Publication 11, 203–23.

Church, M. (1978). Palaeohydrological reconstructions from a Holocene valley fill, in A. D. Miall, (ed.), *Fluvial Sedimentology*, Canadian Society of Petroleum Geologists, Memoir 5, 743–72.

Costa, J. E. (1983). Palaeohydraulic reconstruction of flash flood peaks from boulder deposits in the Colorado Front Range, *Geological Society of America: Bulletin*, **94**, 986–1004.

Costello, W. R., and Walker, R. G. (1972). Pleistocene sedimentology, Credit River, Southern Ontario: A new component of the braided river model, *Journal of Sedimentary Petrology*, **42**, 389–400.

Dawson, M. R. (1985). Environmental reconstructions of a late-Devensian terrace sequence. Some preliminary findings, *Earth Surface Processes and Landforms*, **10**, 237–46.

Dawson, M. R. (1986). *Late Devensian Fluvial Environments of the Lower Severn Basin, UK*, Unpublished Ph.D. Thesis, University of Leicester, 556 pp.

Dawson, M. R., and Bryant, I. D. (1987). Three dimensional facies geometry in Pleistocene outwash sediments, Worcestershire, UK, in F. G. Ethridge, R. M. Flores and M. D. Harvey, (eds), *Recent Developments in Fluvial Sedimentology*, Society of Economic Mineralogists and Paleontologists, Special Publication 39, 191–6.

Dawson, M. R., and Gardiner, V. O. (1987). River terraces: The general model and a palaeohydrological and sedimentological interpretation of the terraces of the lower-Severn, in K. J. Gregory, J. Lewin and J. B. Thornes, (eds), *Palaeohydrology in Practice*, Wiley, London, pp. 271–308.

Eynon, G., and Walker, R. G. (1974). Facies relationships in Pleistocene outwash gravels: A model for bar growth in braided rivers, *Sedimentology*, **21**, 43–70.

Frazer, J. Z. (1982). Derivation of a summary facies sequence based on Markov chain analysis of the Caledon outwash, a Pleistocene fluvial deposit, in R. Davidson-Arnott,

W. Nickling and B. D. Fahey, (eds), *Research in Glacial, Glacio-Fluvial, and Glacio-Lacustrine Systems*, Proceedings of the 6th Guelph Symposium. Geobooks, Norwich, pp. 175–202.

Hey, R. W. (1958). High level gravels in and near the lower Severn Valley, *Geological Magazines,* **95**, 161–8.

Jonnsson, J. (1982). Notes on the Katla volcanological debris flows, *Jokull,* **32**, 61–8.

Maizels, J. K. (1983a). Palaeovelocity and palaeodischarge determination for coarse gravel deposits, in K. J. Gregory (ed.), *Background to Palaeohydrology*, Wiley, Chichester, 101–40.

Maizels, J. K. (1983b) Proglacial channel systems: Change and thresholds for change over long, intermediate and short timescales, in J. D. Collinson and J. Lewin, (eds), *Modern and Ancient Fluvial Systems: Sedimentology and Processes*, International Association of Sedimentologists, Special Publication 6, 251–66.

Maizels, J. K., (1987). Large scale flood deposits associated with the formation of coarse-grained, braided terrace sequences, in F. G. Ethridge, R. M. Flores and M. D. Harvey, (eds), *Recent Developments in Fluvial Sedimentology*, Society of Economic Mineralogists and Paleontologists, Special Publication 39, 135–48.

Reid, I., and Frostick, L. E. (1987). Towards a better understanding of bedload transport, in F. G. Ethridge, R. M. Flores and M. D. Harvey, (eds), *Recent Developments in Fluvial Sedimentology*, Society of Economic Mineralogists and Paleontologists, Special Publication 39, 13–20.

Rust, B. R. (1984). Proximal braidplain deposits in the Middle Devonian Malbaie Formation of Eastern Caspe, Quebec, Canada, *Sedimentology,* **31**, 675–96.

Shotton, F. W. (1977). *The English Midlands*, INQUA Excursion Guide A2, Xth INQUA Congress, Birmingham, 51 pp.

Shotton, F. W., and Coope, G. R. (1983). Exposures in the power house terrace of the River Stour, Wilden, Worcestershire, England, *Proceedings of the Geologists' Association,* **94**, 33–44.

Smith, N. D. (1974) Sedimentology and bar formation in the upper Kicking Horse River: A braided meltwater stream, *Journal of Geology,* **82**, 205–223.

Steel, R. J., and Thompson, D. B. (1983). Structures and textures in Triassic braided stream conglomerates in the Sherwood Sandstone Group, North Staffordshire, England, *Sedimentology,* **30**, 341–68.

Thomas, G. S. P., Connaugton, M., and Dackombe, R. V. (1985). Facies variation in a late Pleistocene outwash sandur from the Isle of Man, *Geological Journal,* **20**, 193–213.

Williams, G. P. (1984). Palaeohydrological equations for rivers', in J. E. Costa and P. J. Fleisher, (eds), *Developments and Applications of Geomorphology*, Springer-Verlag, Berlin, pp. 343–67.

Wills, L. J. (1938). The Pleistocene development of the Severn from Bridgnorth to the sea, *Quarterly Journal of the Geological Society of London,* **94**, 161–242.

Floods: Hydrological, Sedimentological and Geomorphological Implications
Edited by K. Beven and P. Carling
© 1989 John Wiley & Sons Ltd

16 Floods in Fluvial Geomorphology

JOHN LEWIN
University College of Wales, Aberystwyth

INTRODUCTION

The role of floods in geomorphology has long been recognised but it remains persistently controversial. This controversy goes back to that between the creationists and the proponents of the biblical flood, so important to the geologists of the nineteenth century and earlier. Actually observing flood effects came later, and remains difficult, partly because of the logistical problems of being in the right place at the right time given the rather improbable chance of a rare event occurring, and because both equipment and observer may not be able to record what is desirable under the hazardous conditions prevailing.

Modern fluvial geomorphology, and a coherent world-view of the role of floods, took off in the 1950s and 1960s especially with the work of L. B. Leopold, M. G. Wolman and J. P. Miller in *Fluvial processes in geomorphology* and in papers like Wolman and Miller (1960) and Wolman and Leopold (1957). This is not to deny, of course, that there were earlier highly valuable contributions covering similar ground, particularly from river engineers concerned with European rivers (for example, on bed sediment transport) or Indian canals (involving regime theory with links between discharges, channel dimensions, sediment transport and canal stability), and indeed from workers in the United States such as G. K. Gilbert and E. W. Lane. Nevertheless, it is important to appreciate that what amounts to a *fluvial processes* school had a highly pervasive influence and that its quality and consistency of view provided a model as to how and what a generation of fluvial geomorphologists should study. Because there are now some rather different ideas developing, based in particular on new data, this review considers some characteristics of the fluvial processes model and seeks to show just how it needs to be amplified and modified.

THE ORTHODOX MODEL

Three characteristics of the fluvial processes school are worthy of some emphasis.

Firstly, there was the considerable initial benefit of the availability of US data for discharge and channel dimensions at a good set of gauging sites covering environments from glacier outwash to subtropical, and from arid to humid environments. Data were available to a lesser extent for suspended sediment, with some information on bed material. Given the availability of data, and the fact that at least many followers of the fluvial processes school had a background training as much in hydrology and statistics as in hydraulics, then it is not surprising that considerable attention was devoted to regression relationships between morphological variables and measures of discharge, using this near-ubiquitous Q or its surrogate, catchment area, as the independent variable. Given the environmental range of the data, then such findings could also be taken as global in their implications.

Secondly, down on the floodplain channel patterns were mostly studied in alluvial rather than bedrock channels and viewed as essentially braided or meandering, with a separate suite of processes associated with each. It is true that experimental work on initially straight channels was undertaken, but these were regarded as uncommon in nature, whilst buried bedrock meanders were examined also by G. H. Dury (as summarized in Dury, 1964a,b; 1965). However, in process terms, a dichotomy in meandering channels was seen between channel (bed material) and overbank (suspension) deposits, whereas with braided streams the dichotomy was between active and inactive reaches. Given the data available as described above, thresholds between the two could be perceived in slope/discharge terms, so that pattern, too, could be explained in terms of the discharge variable.

Finally, the dimensions of channels were associated with flood events of moderate frequency and magnitude, with overbank flows being achieved every one or two years. This became refined (see Dury 1961; 1976) to around 1.58 years on the annual series, or 1.0 year on the partial duration (peaks over a threshold) series. The idea was that a power function transport rate combined with a right-skewed flow frequency distribution would produce a modal 'effective' discharge (sometimes called 'dominant' or 'formative' discharge). As Dury (1977) has written, 'By a kind of osmotic process, reliance on frequent events of modest magnitude seems to have diffused itself through much of geomorphic thought.'

This broad position, involving the assumption that channels were broadly adjusted to frequent and *hydrologically* identifiable events, had considerable utility. Belief in systems that equilibrated in the short-term allowed students to look at functional inter-relationships between discharge and other environmental variables such as slope, soil and drainage network parameters. Even dynamic studies were possible in the few years available to complete a Ph.D., whilst laboratory flume studies and the natural environment could happily be integrated. For design purposes, a single 'dominant discharge' could also be adopted by engineers with full geomorphological backing (Hey, 1975).

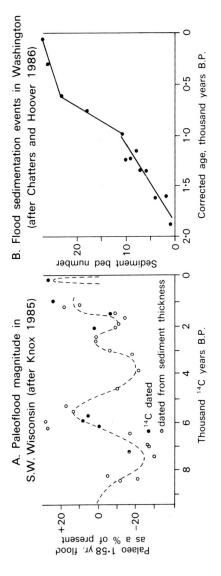

Figure 16.1 Changing flood magnitude, as inferred from the evidence of dated sediments: A, data from palaeochannel dimensions, B, based on the number and data of a sequence of flood-deposited layers.

A final flowering came in palaeohydrological reconstruction where it became evident either through sediment analysis and subsurface exploration (e.g. Dury, 1964b; Rotnicki, 1983), or even by analysis of surface palaeochannels (e.g. Schumm, 1968; Mycielska-Dowgiallo, 1977), that former channels or flows of various sizes were present and that these could be quantitatively linked to former discharges at dated time periods. Two recent examples of work of this kind are illustrated in Figure 16.1. Knox (1985) has analysed sediments and relict channels in south-west Wisconsin to show changes in the magnitudes of 1.58 year floods which could be linked to climatic changes. The study by Chatters and Hoover (1986) again linked climatic fluctuation to variation in dated sedimentary unit thickness, with higher rates of sedimentation and flood magnitudes from 1020 to 1390 A.D.

UNORTHODOX CHALLENGES

Despite its utility, internal consistency and profound influence, the fluvial processes world-view of the role of floods in geomorphology is being changed. There were always exceptional phenomena that extreme floods were known to achieve, such as the transport of very large boulders, whilst slope instability was never fully part of the picture. Challenge to that picture may usefully be examined in four areas—channel patterns, bankfull recurrence interval, flood event observations and basin sediment transport concepts.

Channel patterns

These are undoubtedly more varied than the orthodox view envisaged, and their response to extreme events has proved equally varied. There are thus several mechanisms by which classical braided streams are produced. These include mid-channel deposition and peripheral bar growth (the Leopold and Wolman, 1957, mechanism) but also migratory unit bedforms which may be modified by dissection (e.g. Ashmore, 1982). A separate process of *anastomosis* involves rather stable but multiple sinuous channels which avulse in association with bed aggradation, as in some Canadian rivers (Smith, 1974). Deltaic distributaries would appear to equate with this inland type, though they have been surprisingly less studied. More curiously, high-stage braid formation may occur in conjunction with low-stage anastomosis in Cooper's Creek in Australia, according to Nanson et al. (1986). The floodplains of simple *meandering* rivers are also not generally produced by lateral accretion, with a couplet of coarse bed and fine overbanks deposits, as Leopold and Wolman speculated. Some floodplains of stable meandering streams may be dominated by overbank deposition (so that flows exceeding bankfull stage become more important) whilst deposition of fines *within* channels which cannot be nearly so easily related to channel-filling flow frequencies are also known to be significant (Nanson, 1980; Nanson and Page, 1983). Another category of *wandering* river, involving an irregular sinuous channel sometimes split by islands and braiding, has also been identified (Church, 1983; Desloges and Church, 1987).

Even the much-quoted slope–discharge threshold between braiding and meandering has been reconsidered in apparently independent reappraisals (Ferguson, 1984; Carson, 1984). These studies indicate that because coarse sediment needs higher stream power for transportation, streams with such bed-materials must plot higher on the slope–discharge graph. The plot thus reflects sediment size, and the fact that coarse bedload streams commonly are braided is incidental. High *load* of bed-calibre material is the missing prerequisite for braiding to occur.

A last point concerning channel pattern processes is that the orthodox approach did not assign an adequate role to the process of avulsion. This may be important in producing channel switching which is involved in the reordering of braided or wandering systems (Ferguson and Werritty, 1983) and alluvial fans, but many geologists have seen it also as most important on big rivers as revealed in the development of thick alluvial sequences (Bridge and Leeder, 1979). This observation may be appreciated from both historical and stratigraphic studies: the catastrophic historical changes on the Huang He in China, with disastrous loss of life, and the episodic Holocene relocation of Mississippi Delta lobes, appear both to result from flood avulsion and testify to the importance of rare events in some alluvial environments where slow accretion (in levées, channels or alluvial ridges) may be an essential prelude to episodic abandonment of channels that have become progressively elevated. Such environments may be termed ones of 'accretion catastrophe'.

Bankfull recurrence interval and dominant discharge

There have been persistent suggestions that channel dimensions do not universally and consistently identify with the most probable annual flood. Figure 16.2 plots data from six studies; some show close identification between channel dimensions and frequent events, whilst others show much more variety. In part this could be the result of channel incision at some sites, but more systematic reasons have been suggested. Thus Harvey (1969) showed that three small streams behaved rather differently, a base-flow dominated chalk stream having a bankfull discharge more rarely than the others. Pickup and Warner (1976) suggested that rarer bankfull floods determined channel capacity whereas smaller ones were the most effective in transporting bedload. More in line with the orthodox view, Andrews (1980) and Webb and Walling (1982) suggested that flows at around bankfull were indeed the most important single flow class for mechanical river load transport (see Figure 16.3A). Such sediment data are very rarely available for the many years needed to accommodate rare extreme events, however.

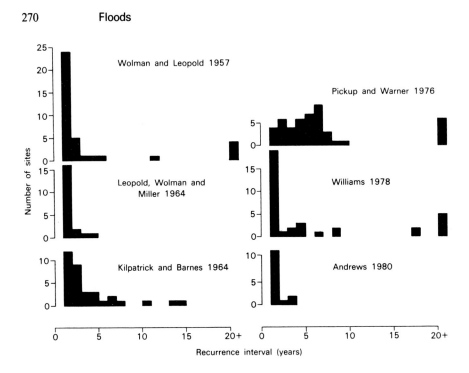

Figure 16.2 Selected studies of the recurrence interval of bankfull discharge.

Some interest has been shown in examining channel-full discharges within single basins (Figure 16.4). Dury (1961) reported that the duration of overbank discharges could increase down-river, whilst Wolman (1955) suggested tentatively that bankfull recurrence interval, too, might decrease downstream (Figure 16.4A). Far more extreme results were reported by Nanson and Young (1981) with first an increase and then a considerable decrease in bankfull channel capacity in the downstream direction (Figure 16.4B). Figure 16.4C shows estimated channel-full discharge for the River Vyrnwy in the 20 km above its junction with the River Severn. This reach is characterised by marked changes in channel mobility, gradient and bank materials from a laterally mobile steeper-gradient and coarser bed-material zone upstream to a stable, very gentle gradient channel in finer sediment downstream. The confluence area is known as being flood-prone, and although there are difficulties inherent in estimating discharges for numerous cross sections, this does appear to be a case of greater use of overbank flow conveyance downstream.

It is worth digressing at this point on possible approaches to be adopted in this kind of work. Firstly, the regression exercises based on sound gauging station data, although referred to under the heading of 'downstream hydraulic

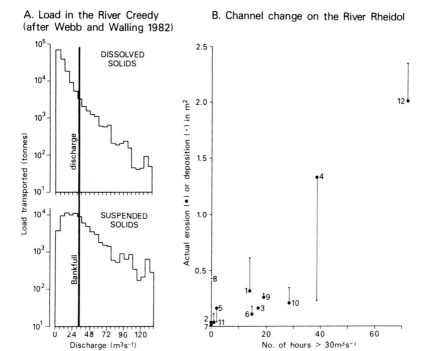

Figure 16.3 (A) Load transported by varying discharge classes in relation to bankfull discharge on the River Creedy, Devon; (B) channel change in relation to duration of flows above a threshold of $30\,\mathrm{m^3\,s^{-1}}$, River Rheidol, Wales.

geometry', were so-called in a rather particular sense. They usually do not involve downstream changes on single streams (as in Figure 16.4); the 'geometry' identified by Leopold and Maddock (1953) was also not *channel* geometry but the 'geometric' relationship between variables (straight-line fits on log-log plots), whilst hydraulics were involved only in the sense that the variables (width, depth, velocity, slope and roughness) were to do with flow in channels.

Taken further, correlation techniques have been used to identify most effective discharges. For example Carlston (1965) examined relationships between meander dimensions and a range of discharge parameters, those showing the best correlations being taken as the most likely to be physically the most important causative ones.

Other more truly hydraulic approaches are possible, as for example Ferguson (1986) has suggested with regard to 'hydraulic geometry'. It is possible to observe physical mechanisms on-site to assess directly what conditions are responsible for what. Identifying a single class of flows that are the *most* effective may be convenient, but a whole range of flows can in reality be important (Harvey *et al.*,

272   Floods

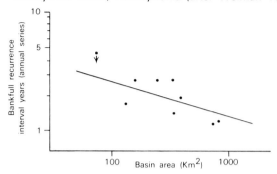

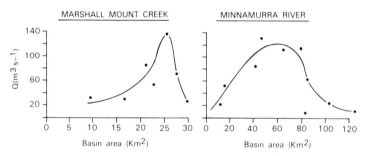

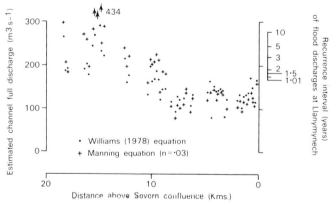

Figure 16.4 Bankfull discharge within single drainage basins in relation to drainage area or downvalley distance. The Williams (1978) equation relates bankfull discharge (Q_b) to channel cross-sectional area (A_b) and slopes (S) as $Q_b = 4.0\, A_b^{1.21} S^{0.28}$.

1979), as in Figure 16.3A, in which above-bankfull flows are important for suspended sediment transport as well as ones at and well below. Figure 16.3B, concerned with bank erosion, shows another approach: a discharge of $30 \, m^3 \, s^{-1}$ (just under half the mean annual flood figure) appeared from field observations, at the illustrated cross section, to be the threshold for bed sediment transport and for each event during a winter, the *duration* above that threshold appears to relate well to bank erosion rate—better, in fact than the peak magnitudes of the events involved.

Flood event observations

Recent years have seen a growing number of studies of specific events (e.g. Gupta, 1983). Many of these appear to produce exceptional landscape features, but some it is possible to set within a record of sediment yield or work accomplished by more normal events (see Figure 16.5). In the case of Hurricane Agnes in June 1982, recovery and reversion to 'normal' conditions appear to have been quite rapid in the north-eastern United States (Costa, 1974; Gupta and Fox, 1974; Dury, 1977). Elsewhere flood effects appear to have been dramatic, as in Coffee Creek, California (Stewart and La Marche, 1967), Eldorado Canyon, Nevada (Glancy and Harrison, 1975) and the Big Thompson Canyon in Colorado (Shroba *et al.*, 1979). Dam bursts which may enhance or produce floods have also been studied in some detail (Scott and Gravlee, 1968; Costa, 1985).

There appears to be a growing impression that some environments are more prone to the long-lasting landforming effects of exceptional events than others. For example, dramatic channel changes in response to floods, or sets of floods, have been reported for semi-arid environments (Schumm and Lichty, 1963; Burkham, 1972). Forms produced here by floods seem also to persist, though they may be localized (Baker, 1977; Thornes 1976; Harvey, 1984), to a greater extent than some humid areas (see Anderson and Calver, 1977). Other environments may be persistently affected by extreme events as well. On the Eel River in California (Figure 16.5) considerable changes occurred following an extreme flood which triggered large numbers of landslides (Brown and Ritter, 1971; Cleveland, 1977). Stream bed elevations were considerably raised (paradoxically reducing bank erosion subsequently) returning slowly to pre-1964 conditions.

This leads to a second point which concerns the considerable growing importance ascribed to the triggering of slope instability, the significance of this in small headwater catchments rather than larger basins, and the subsequent transmission of such sediments through basins in the post-event period (e.g. Caine, 1980; Cleveland, 1977; Newson, 1980; Starkel, 1976; Tanakam 1976; Wells and Harvey, 1987; Wolman and Gerson, 1978). The recurrence interval for major slope instability at a site can be high, but sediments derived from events at sets of sites may be an important antecedent condition for bed sediment supply and

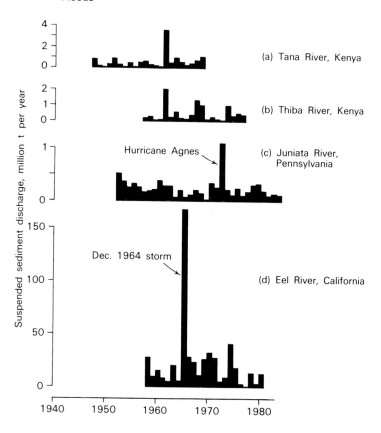

Figure 16.5 Annual suspended sediment loads for selected rivers. (a) and (b) from Walling (1984); (c) and (d) from Meade and Parker (1985).

distribution within catchments by more frequent competent discharges in-channel.

A third point is that there are regional and scale differences in water discharge events themselves. Figure 16.6 plots regional growth curves showing the magnitude of discharges for given return periods in relation to the magnitude of the $Q_{2.33}$ flood using annual flood data for countries which illustrate contrasted world hydrological environments. The varying magnitude of extremes of equal probability in different environments is clear, with monsoon and semi-arid environments (Sri Lanka, south-western United States) producing floods which are much larger in relation to the 2.33-year flood than in tropical environments (Congo and Guyana). Figure 16.7 illustrates also the importance of catchment size as well as environment. Magnitude ratios for the 50-year flood are plotted in relation to catchment area for different parts of the United States. A decrease in

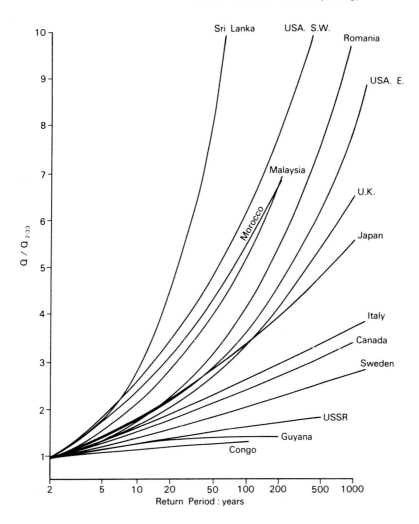

Figure 16.6 Regional growth curves, showing relative flood magnitudes plotted against return period for selected countries. Analysis from Institute of Hydrology (undated); data from UNESCO (1976).

the relative magnitude of the 50-year flood with increasing catchment size is apparent, but also the greater and more variable discharges for smaller catchments in semi-arid New Mexico.

One may conclude from this discussion that analysis of flood data and the observation of geomorphological work accomplished by floods in different catchments points up important contrasts. The relative magnitude of rare events of

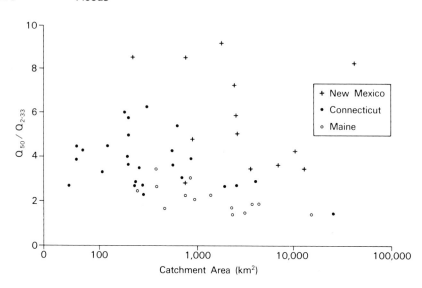

Figure 16.7 Relative flood magnitude and catchment area. Data from Benson (1962; 1964).

given frequency differs according to both hydroclimatic environment and catchment size; infrequent slope instability may be coupled with more frequent channel transport to produce a nested arrangement of effective events that are together responsible for whole-basin sediment outputs.

Basin sediment transport concepts

A final set of factors changing geomorphologists' approaches to flood events concerns changing ideas concerning sediment system modelling, for it is essentially through sediment transport that flood-related landforms are produced. Here the most notable figure concerned with general principles has been S. A. Schumm. Inspired particularly by observations in semi-arid environments and by physical model studies (Schumm *et al.*, 1987), Schumm has championed a number of concepts, particularly ones involving geomorphic *thresholds*, the necessity of considering catchment systems as a whole, and the possibility in particular of catchment *complex response* to forcing events (Schumm, 1977). Others have seen catchments as variably *sensitive* to such events (Brunsden and Thornes, 1979), and there is a growing interest in routing sediment through systems (e.g. Dietrich and Dunne, 1978; Meade, 1982) which may have spatially variable potential for sediment storage and the transmission or damping of flood effects. A second and related set of useful ideas has concerned the response and recovery times to extreme events. These have been generally reviewed by Wolman and Gerson (1978), again pointing to spatial and temporal variability in

Floods in Fluvial Geomorphology

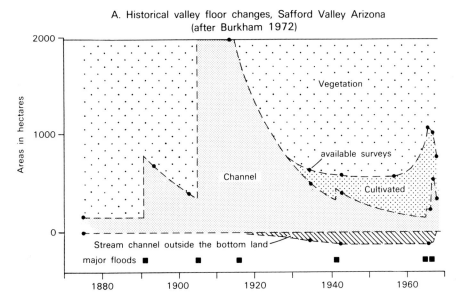

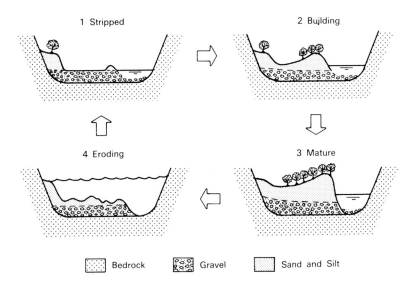

Figure 16.8 Geomorphological changes on floodplains in relation to channel metamorphosis and valley-floor stripping.

the recovery of systems in the face of extreme events. The idea that alluvial channels may follow long-term histories of triggering and recovery may be appreciated from Figure 6.8, which summarizes work by Burkham (1972) and Nanson (1986). In both cases the persistent effect of major transformations lasting for decades, or in the case of streams studies by Nanson, hundreds or thousands of years, is emphasised. Similar concentration on important extreme events, though ones way beyond those recorded in the comparatively short runs of historical records, has been the focus of work by V. R. Baker and his students (e.g. Baker, 1973; Patton *et al.*, 1979; Baker and Pickup, 1987). Their work involves the use of dated flood sediments as records of flood stages, particularly in critical topographic and climatic environments where such sediments may be preserved (notably bedrock canyons in semi-arid environments allowing the preservation of slack-water sediments). Finally, on the reach scale, there has been a considerable amount of research on the flow conditions (shear stress, velocity or stream power) necessary to initiate and maintain sediment transport. In particular, relationships between sediment size and transporting capacity have been sought, often in an endeavour to reconstruct palaeo-discharges (Church, 1978; Bradley and Mears, 1980; Costa, 1983; Maizels, 1983; Williams, 1983). The researchers involved view their results with considerable caution. Coarse sediment may be transported in the form of debris flow (Costa and Jarrett, 1981), and actual relationships between flows and sediment transport are proving to be more complex in natural streams than might have been thought (e.g. Reid *et al.*, 1985). In large part this arises because of mixed sediment size populations, involving the 'hiding' of small particles by larger ones, bed armouring and imbrication, and greater exposure of isolated large particles than would be the case with uniform bed materials of the same size.

DISCUSSION

Some decades of research have now shown (and it would indeed be a matter of surprise and regret if they had not, and is no reflection on the pioneer authors earlier cited) that the 'orthodox' model presented earlier is in need of revision and amplification. It should not be taken as a universal world model for geomorphology. Slightly amplifying some conclusions of Baker (1977), we should expect the role of extreme floods to be geographically varied, particularly because of three sets of factors:

(1) The relative magnitude of extreme floods, and thus the potential for the incidence and preservation of 'catastrophic landforms', varies. It is greatest in certain climatic environments and probably in smaller catchments. As emphasised by Wolman and Gerson (1978), magnitude–frequency effects vary with scale and environment.

(2) Physiographic factors are significant in that they involve flood-flow generation, contrasted slope instability conditions, the potential for preserving the effects of floods selectively, and different sedimentary environments which may respond to extreme flows in a variable manner. Basin systems may usefully be looked at in terms of variable response in their different parts.

(3) Sediments have generally to be considered more systematically. Thus the *record* of extreme events often depends on having large boulders *available* for entrainment. The hydraulics of sediment transport equally cannot be considered in fluid force terms alone without incorporating sediment parameters as well. Rather than a linear response to discharge, increasing flood discharge may encounter a series of nested *thresholds* (e.g. for sediment entrainment or slope stability) so that stepped relationships could be more appropriate.

Such factors make the study of floods in fluvial geomorphology more difficult even than they once appeared. Away from gauging stations, working in basin systems, data for discharge may become less reliable, whilst the dearth of sediment-related information can only be remedied with difficulty and persistence. In future, however, it may perhaps be helpful to appreciate that research needs to be conducted in at least five rather contrasted geomorphological settings:

(a) In *headwater environments*, extreme events and slope instability may couple to produce notably impressive effects on a localised basis: such effects may be dampened out in larger basins during the event concerned, but may be an essential ingredient of average sediment supply in the longer term as such materials are transferred from individual sites into larger drainage basins.

(b) *Bedrock gorges* are amongst a group of environments that may be distinctive in preserving the effects of high-magnitude events, as shown by the work of V. R. Baker and others, and these may be the events to which major landforms relate. This can apply elsewhere, as in alluvial fan environments (Beaty, 1974) where events that are rare in human terms may be responsible for normal fan construction, with very little activity in between times.

(c) *Channel metamorphosis* and recovery in alluvial rivers is characteristic of some climatic and hydraulic environments. Here as both individual studies have documented (Figure 16.8), and as broader theoretical ideas have indicated, some floodplains as a whole are 'catastrophe-related' in that their morphology is best interpreted as a response to and recovery from rare events, and ones which short runs of hydrological records may not incorporate.

(d) *Lateral accretion environments,* as characterised by the middle reaches of many river systems and many humid-temperate environments, may still approximate to many features of the 'orthodox' model. Development of ideas on channel pattern formation, on sediment transport and on the effectiveness of flows with different recurrence intervals has and will occur but many of the pioneer ideas of the fluvial processes school do retain support from studies that have been undertaken subsequently.

(e) Finally, there are the *accretion catastrophe* environments of large rivers which require greater attention. Here avulsion following large floods may catastrophically relocate channels; there is also a range of anastomosing deltaic and other channel patterns to be further studied in terms of their response to events that may not necessarily be very extreme, but are rare in that a very long period of prior preparation is necessary beforehand. In particular attack on large-scale problems, both academic and practical, is required to supplement and extend the older research in mid-latitudes, on a scale from models to moderate-sized basins, which has proved so important.

REFERENCES

Anderson, M. G., and Calver, A. (1977). On the persistence of landscape features formed by a large flood. *Institute of British Geographers Transactions*, NS2, 243-54.

Andrews, E. D. (1980). Effective and bankfull discharges of streams in the Yampa river basin, Colorado and Wyoming, *Journal of Hydrology*, 46, 311-30.

Ashmore, P. E. (1982). Laboratory modelling of gravel braided stream morphology, *Earth Surface Processes and Landforms*, 7, 201-25.

Baker, V. R. (1973). *Paleohydrology and Sedimentology of Lake Missoula Flooding in Eastern Washington*, Special Paper, Geological Society of America, no. 144, 79 pp.

Baker, V. R. (1977). Stream-channel response to floods, with examples from central Texas. *Geological Society of America Bulletin*, 88, 1057-71.

Baker, V. R., and Pickup, G. (1987). Flood geomorphology of Katherine Gorge, Northern Territory, Australia, *Geological Society of America Bulletin*, 98, 635-46.

Beaty, C. B. (1974). Debris flows, alluvial fans and a revitalized catastrophism, *Zeitschrift fur Geomorphologie*, Suppl. 21, 39-51.

Benson, M. A. (1962). *Factors Influencing the Occurrence of Floods in a Humid Region of Diverse Terrain*, United States Geological Survey Water Supply Paper, 1580-B, 64 pp.

Benson, M. A. (1964). *Factors Affecting the Occurrence of Floods in the Southwest*, United States Geological Survey Water Supply Paper, 1580-D, 72 pp.

Bradley, W. C., and Mears, A. I. (1980). Calculation of flows needed to transport coarse fraction of Boulder Creek alluvium at Boulder, Colorado, *Geological Society of America Bulletin*, 91, 1056-90.

Bridge, J. S., and Leeder, M. R. (1979). A simulation model of alluvial stratigraphy, *Sedimentology*, 26, 617-44.

Brown, W. A. III, and Ritter, J. R. (1971). *Sediment Transport and Turbidity in the Eel River Basin, California*, United States Geological Water Supply Paper, 1986, 70 pp.

Brunsden, D., and Thornes, J. B. (1979). Landscape sensitivity and change. *Institute of British Geographers Transactions*, **NS4**, 463–84.

Burkham, D. E. (1972). *Channel Changes on the Gila River in Safford Valley, Arizona, 1846–1970*. United States Geological Survey Professional Paper, 655–G, 24 pp.

Caine, T. N. 1980. The rainfall intensity-duration control of shallow landslides and debris flows, *Geografiska Annaler*, **62A**, 23–7.

Carlston, C. W. (1965). The relation of free meander geometry to stream discharge and its geomorphic implications, *American Journal of Science*, **263**, 864–85.

Carson, M. A. (1984). The meandering-braided river threshold: a reappraisal, *Journal of Hydrology*, **73**, 315–34.

Chatters, J. C., and Hoover, K. A. (1986). Changing Late-Holocene flooding frequencies in the Colombia River, Washington, *Quaternary Research*, **26**, 309–20.

Church, M. (1978). Palaeohydrological reconstructions from a Holocene valley fill, in A. D. Miall, (ed.), *Fluvial Sedimentology*, Canadian Society of Petroleum Geologists, Calgary, 743–72.

Church, M. (1983). Pattern of instability in a wandering gravel bed channel, in J. D. Collinson and J. Lewin, (eds), *Modern and Ancient Fluvial Systems*, Special publ. no. 6, International Association of Sedimentologists, Blackwell, Oxford, pp. 169–80.

Cleveland, G. B. (1977). Rapid erosion along the Eel River, California, *California Geology*, **30**, 204–11.

Costa, J. E. (1974). Response and recovery of a Piedmont watershed from tropical storm Agnes, June 1972, *Water Resources Research*, **10**, 106–12.

Costa, J. E. (1983). Paleohydraulic reconstruction of flash-flood peaks from boulder deposits in the Colorado Front Range. *Geological Society of America Bulletin*, **94**, 986–1004.

Costa, J. E. (1985). *Floods from Dam Failures*, United States Geological Survey Open File Report, 85–560, 85–560, 84 pp.

Costa, J. E., and Jarrett, R. D. (1981). Debris flow in small mountain stream channels of Colorado and their hydrological implications, *Bulletin of the Association of Engineering Geologists*, **18**, 309–22.

Desloges, J. R., and Church, M. (1987). Channel and floodplain facies in a wandering gravel-bed river, *Society of Economic Paleontologists and Mineralogists*, Special publication no. 39, 99–109.

Dietrich, W. E., and Dunne, T. (1978). Sediment budget for a small catchment in mountainous terrain, *Zeitschrift für Geomorphologie*, **NF 29**, 191–206.

Dury, G. H. (1961). Bankfull discharge: an example of its statistical relationships, *International Association of Scientific Hydrology*, Publ. no. 3, 48–55.

Dury, G. H. (1964a). *Principles of Underfit Streams*, United States Geological Survey Professional Paper, 452–B, 67 pp.

Dury, G. H. (1964b). *Subsurface Explorations and Chronology of Underfit Streams*, United States Geological Survey Professional Paper, 452–B, 56 pp.

Dury, G. H. (1965). *Theoretical Implications of Underfit Streams*, United States Geological Survey Professional Paper, 452–C, 43 pp.

Dury, G. H. (1976). Discharge prediction, present and former, from channel dimensions, *Journal of Hydrology*, **30**, 219–45.

Dury, G. H. (1977). Peak flows, low flows and aspects of geomorphic dominance, in K. J. Gregory, (ed.), *River Channel Changes*, Wiley, Chichester, pp. 61–74.

Ferguson, R. I. (1984). The threshold between meandering and braiding, in K. V. H. Smith, (ed.), *Channels and Channel Control Structures*, Springer-Verlag, Berlin, pp. 6.15–6.29.

Ferguson, R. I. (1986). Hydraulics and hydraulic geometry, *Progress in Physical Geography*, **10**, 1–31.

Ferguson, R. I., and Werritty, A. (1983). Bar development and channel change in the gravelly River Feshie, in J. D. Collinson and J. Lewin, (eds), *Modern and Ancient Fluvial Systems*, Special publ. no. 6, International Association of Sedimentologists, Blackwell, Oxford, pp. 181–93.

Glancy, P. A., and Harrison, L. (1975). *A Hydrologic Assessment of the September 14, 1974, Flood in Eldorado Canyon, Nevada*, United States Geological Survey Professional Paper, 930, 28 pp.

Gupta, A. (1983). High magnitude floods and stream channel response, in J. D. Collinson and J. Lewin, (eds) *Modern and Ancient Fluvial Systems*, Special publ. no. 6, International Association of Sedimentologists, Blackwell, Oxford, pp. 219–227.

Gupta, A., and Fox, A. (1974). Effects of high-magnitude floods on channel form: a case study in Maryland Piedmont. *Water Resources Research*, **10**, 499–509.

Harvey, A. M. (1969). Channel capacity and the adjustment of streams to hydrological regime, *Journal of Hydrology*, **8**, 82–98.

Harvey, A. M. (1984). Geomorphological response to an extreme flood: a case from southeast Spain. *Earth Surfaces Processes and Landforms*, **9**, 267–79.

Harvey, A. M., Hitchcock, D. H., and Hughes, D. J. (1979). Event frequency and morphological adjustment of fluvial systems in upland Britain, in D. D. Rhodes and G. P. Williams, (eds), *Adjustments of the Fluvial System*, Kendal-Hunt, Dubuque, Iowa, pp. 139–67.

Hey, R. D. (1975). Design discharges for natural channels, in R. D. Hey and T. D. Davies, (eds), *Science, Technology and Environmental Management*, Saxon House, Farnborough, pp. 73–88.

Institute of Hydrology (no date) *Research Report 1981–84*, Natural Environment Research Council, 86 pp.

Kilpatrick, F. A., and Barnes, H. H. (1964). *Channel Geometry of Piedmont Streams as Related to Frequency of Flooding*, United States Geological Survey Professional Paper, 422–E, 10 pp.

Knox, J. C. (1985). Responses of floods to Holocene climatic change in the Upper Mississippi valley, *Quaternary Research*, **23**, 287–300.

Lane, E. W. (1957). *A Study of the Shape of Channels Formed by Natural Streams Flowing in Erodable Materials*. MRD Sediment Series no. 9.

Leopold, L. B., and Maddock, T. Jr. (1953). *The Hydraulic Geometry of Stream Channels and Some Physiographic Implications*, United States Geological Professional Paper, 252, 57 pp.

Leopold, L. B., and Wolman, M. G. (1975). *River Channel Patterns: Braided Meandering and Straight*, United States Geological Survey Professional Paper, 282–B, 85 pp.

Leopold, L. B., Wolman, M. G., and Miller, J. P. (1964). *Fluvial Processes in Geomorphology*, Freeman, San Francisco, 522 pp.

Maizels, J. K. (1983). Palaeovelocity and palaeodischarge determination for coarse gravel deposits, in K. J. Gregory, (ed.), *Background to Palaeohydrology*, Wiley, Chichester, pp. 101–39.

Meade, R. H. (1982). Sources, sinks and storage of river sediment in the Atlantic drainage of the United States, *Journal of Geology*, **90**, 235–52.

Meade, R. H., and Parker, R. S. (1985). *Sediment in Rivers of the United States*. United States Geological Survey Water Supply Paper, 2275, 49–60.

Mycielska-Dowgiallo (1977). Channel pattern changes during the Last Glaciation and Holocene, in the northern part of the Sandomiez Basin and the middle part of the Vistula Valley, Poland, in K. J. Gregory, (ed.), *River Channel Changes*, Wiley, Chichester, pp. 75–87.

Nanson, G. C. (1980). Point bar and floodplain formation on the meandering Beatton River, northeastern British Columbia, Canada, *Sedimentology*, **27**, 3–29.

Nanson, G. C. (1986). Episodes of vertical accretion and catastrophic stripping: a model of disequilibrium floodplain development, *Geological Society of America Bulletin,* **97**, 1467-75.

Nanson, G. C., and Page, K. (1983). Lateral accretion of fine-grained concave benches on meandering rivers, in J. D. Collinson, and J. Lewin, (eds), *Modern and Ancient Fluvial Systems*, International Association of Sedimentologists, Special publ. no. 6, Blackwell, Oxford, pp. 133-45.

Nanson, G. C., Rust, B. R., and Taylor, G. (1986). Coexistent mud braids and anastomosing channels in an arid-zone river: Cooper Creek, central Australia, *Geology,* **14**, 175-8.

Nanson, G. C., and Young, R. W. (1981). Downstream rduction of rural channel size with contrasting urban effects in small coastal streams of southeastern Australia, *Journal of Hydrology,* **52**, 239-55.

Newson, M. (1980). The geomorphological effectiveness of floods—a contribution stimulated by two recent events in mid-Wales, *Earth Surface Processes,* **5**, 1-16.

Patton, P. C., Baker, V. R., and Kochel, R. C. (1979). Slackwater deposits: a geomorphic technique for the interpretation of fluvial paleohydrology, in D. D. Rhodes and G. P. Williams, (eds), *Adjustments of the Fluvial System*, Kendal-Hunt, Dubuque, Iowa, pp. 225-52.

Pickup, G. and Warner, R. F. (1976). Effects of hydrological regime on magnitude and frequency of dominant discharge, *Journal of Hydrology,* **29**, 51-75.

Reid, I., Frostick, L. E., and Layman, J. T. (1985). The incidence and nature of bedload transport during flood flows in coarse-grained alluvial channels, *Earth Surface Processes and Landforms,* **10**, 33-44.

Rotnicki, K. (1983). Modelling past discharges of meandering rivers, in K. J. Gregory, (ed.), *Background to Palaeohydrology*, Wiley, Chichester, pp. 322-54.

Schumm, S. A. (1968). *River Adjustment to Altered Hydrologic Regimen: Murrumbiggee River and Palaeochannels, Australia.* United States Geological Survey Professional Paper, 598.

Schumm, S. A. (1977). *The Fluvial System*, Wiley, New York, 338 pp.

Schumm, S. A., and Lichty, R. W. (1963). *Channel Widening and Flood-plain Construction Along Cimarron River in Southwestern Kansas,* United States Geological Survey Professional Paper, 352-D, 71-88.

Schumm, S. A., Mosley, M. P., and Weaver, W. E. (1987). *Experimental Fluvial Geomorphology*, Wiley, New York, 413 pp.

Scott, K. M., and Gravlee, G. C. (1968). *Flood Surge on the Rubicon River*, California— Hydrology, Hydraulics and Boulder Transport. United States Geological Survey Professional Paper, 422-M, 40 pp.

Shroba, R. R., Schmidt, P. W., Crosby, E. J., and Hansen, W. R. (1979). *Storm and Flood of July 31–August 1, 1976, in the Big Thompson and Cache La Pondre River Basins, Larimer and Weld Counties, Colorado,* United States Geological Survey Professional Paper, 1115, Part B, 87-152.

Smith, D. G. (1974). Aggradation of the Alexandra-North Saskatchewan River, Banff Park, Alberta, in M. E. Morisawa, (ed.), *Fluvial Geomorphology*, State University of New York, Binghamton, pp. 201-9.

Starkel, L. 1976. The role of extreme (catastrophic) meteorological events in contemporary evolution of slopes, in E. Derbyshire, (ed.), *Geomorphology and Climate*, Wiley, Chichester, pp. 203-46.

Stewart, J. H., and La Marche, V. C. (1967). *Erosion and Deposition Produced by the Flood of December 1964 on Coffee Creek, Trinity County, California.* United States Geological Survey Professional Paper, 422-K, 22 pp.

Tanaka, M. (1976). Rate of erosion in the Tanzawa mountains, central Japan,

Geografiska Annaler, **58A**, 155-64.
Thornes, J. B. (1976). *Semi-arid Erosion Systems: Case Studies from Spain,* London School of Economics, Geographical Papers, no. 7, 79 pp.
UNESCO (1976). *World Catalogue of Very Large Floods,* UNESCO, Paris, 424 pp.
Walling, D. E. (1984). The sediment yield of African rivers. *IASH publication no. 144,* 265-83.
Webb, B. W., and Walling, D. E. (1982). The magnitude and frequency characteristics of fluvial transport in a Devon drainage basin and its geomorphological implications, *Catena,* **9**, 9-23.
Wells, S. G., and Harvey, A. M. (1987). Sedimentological and geomorphic variations in storm-generated alluvial fans, Howgill Fells, northwest England, *Geological Society of America Bulletin,* **98**, 198.
Williams, G. P. (1978). Bank-full discharge of rivers, *Water Resources Research,* **14**, 1141-54.
Williams, G. P. (1983). Paleohydrological methods and some examples from Swedish fluvial environments I Cobble and boulder deposits, *Geografiska Annaler,* **65A**, 229-43.
Wolman, M. G. (1955). *The Natural Channel of Brandywine Creek, Pennsylvania,* United States Geological Survey Professional Paper, 271, 56 pp.
Wolman, M. G., and Gerson, R. (1978). Relative scales of time and effectiveness of climate in watershed geomorphology, *Earth Surface Processes,* **3**, 189-208.
Wolman, M. G., and Leopold, L. B. (1957). *River Floodplains: Some Observations on their Formation,* United States Geological Survey Professional Paper, 282-C, 109 pp.
Wolman, M. G., and Miller, J. P. (1960). Magnitude and frequency of forces in geomorphic processes, *Journal of Geology,* **68**, 54-74.

INDEX

A horizon 193, 194, 195
Accelerator mass spectrometry 175
Accretion 162, 166, 280
  catastrophe 269, 280
Aeolian material 179
Aggradation 187, 240
Allochthonous 175
Alluvial
  channels 266
  fans 234, 279
  landforms 233
  rivers 171
Alluvium 187, 253
Aluminium 195
Anabranches 255
Anastomosis 195, 268, 280
Ancient floods 107, 128, 171, 175, 178,
    181, 248, 261, 268
Annual flood series 171, 266, 274
Antecedent conditions 229, 273
Ants 189
Archaeology 175
Architecture 254
Arid climates 172, 174
Armouring 144, 278
Ash deposits 239
Atlantic period 229
Autochthonous 175
Avulsion 233, 234, 254, 269, 280

B horizon 193, 194
Backwater 176, 189
Bank erosion 233, 273
Bank undercutting 207
Bankfull discharge 3, 45, 78, 171, 270
Bars 144, 193, 205, 207, 212, 224, 225
Base level 234, 256

Baseflow 269
Basin morphology 161
Bed slope 261
Bedforms 135, 144, 258, 259, 268
Bedload 119, 201, 207
  pulses 258
  traps 109
Bedrock 172
  canyons 278
  channels 266
  meanders 266
Bella Coola River 163
Benchmarks 195
Berm 186
Big Thompson Canyon 273
Bingham behaviour 242
Bioturbation 174
Bog
  burst 155
  failure 199
  flow 155
  slide 155, 201, 205
Boulder
  bars 144, 193, 205, 207, 212, 224, 225
  berm 186
  jam 205, 215
Boulder Creek, Colorado 107
Boundary shear stress 93
Braided channels 200, 209, 225, 234, 254,
    260, 268
Braiding index 224, 225, 228, 233
Breccia 240
Brown earth 193, 194
Bulk density 242
Bulk strength 248
Buoyancy 242
Buried soils 175

C horizon 193
Caldwell Burn 158
California 115, 129
Cambourne 60
Canal stability 265
Canyon 273, 278
Carboniferous 199
Carl Beck 125, 126
Catastrophic floods 8, 128, 151, 155, 172, 229, 239, 268, 278
Catastrophists 161
Catchment characteristics 65, 214
Censoring 176, 179, 180
Chalk streams 77, 269
Channel
   adjustment 162, 219
   capacity 67, 78, 99, 270
   enlargement ratio 70
   erosion 69, 70, 269
   expansion 172
   improvement 38, 57, 62
   incision 269
   metamorphosis 279
   morphology 62
   order 158
   pattern 220, 268
   planform 220, 225
   sedimentation 69
   switching 269
Charcoal 175
Chippenham 60
Chronology 195
Chute 207, 215
Clay 159, 200, 242, 245, 247
   drapes 257, 258, 259, 260
Climatic change 152, 155, 220, 228, 233 ff, 268
Climbing ripples 245
Cloudburst 1, 193
Cobbles 164
Coffee Creek 273
Colluvium 174, 179
Competence 107 ff, 212, 215, 274
Complex response 276
Confluence 135, 270
Confluences 135
Convectional storm 142, 144, 202
County Leitrim 199
County Sligo 199
Creationists 265
Cross-bedding 245, 247, 256 ff

Cumbernauld 60
Cut-off 205
Cyclic grading 247
Cyclical history 219, 247
Cyclonic storms 229

Dam bursts 52, 152, 249, 273
Dating 175, 186, 196, 249, 278
Debris fan 205
Debris flows 7, 159, 161, 239, 242, 278
   cohesive 242
Debris torrent 159
Deforestation 229, 234, 235
Delivery ratio 163
Deltaic 268, 280
Dendrochronology 175
Deposition 195, 222
Devensian 187, 189, 253
Dewatering 248
Diamicton 200, 253
Diffusion wave 52
Dimlington 189, 194
Discharge ratio 139, 141
Dispersive stress 242, 247
Distributary fans 225
Doheny Channel, California 130
Dolgarrog 152, 159
Dominant discharge 266
Drapes 258 ff
Drift 161, 187, 202, 204
Drought 152, 155
DuBoys equation 2, 99, 129
Dunes 247, 248

Effective discharge 266, 271
Effectiveness 3, 151, 155, 158, 161, 163, 166
Eldorado Canyon 273
Energy slope 161
Entrainment 108
Ephemeral channels 138
Exceedance 180
Extreme floods 158, 161, 179, 181, 195, 199, 220, 229, 233, 234, 239, 268, 280
Extrinsic threshold 233

Fan lobe 248, 249
Fining-upward deposits 245, 247
Flood
   chronology 228
   control 47

Index 287

Flood (*cont.*)
  deposits 161, 185, 189
  depth 155, 161, 177, 194
  extreme 158, 161, 179, 181, 195, 199, 220, 229, 233, 234, 239, 268, 280
  flash 220
  frequency 3, 6, 44, 77, 79, 164, 171, 181, 228
  hazard 152
  history 185, 195
  hydrographs 140, 261
  peak magnitude 180, 213, 253
  power 161
  protection 47
  routing 49, 98
  sediments 135
  stage 161, 177, 194
Flood channel facility 101
Flood studies report 11, 37, 65
Flood waves
  attenuation 42
  confluences 136, 140
  diffusion wave 52
  kinematic wave 52
  St Venant equations 52
  travel time 41
  unsteady flow 49, 89
  wave speed 41
Floodplain 45, 69, 162, 166, 195, 209, 212, 253, 279
  channel 266
  stratigraphy 164
  stripping 161
  zoning 179
Floods, ancient 107, 128, 171, 175, 178, 181, 248, 261, 268
Flow
  competence 107 ff
  conveyance 270
  duration 273
  resistance 84, 87, 159, 212
Fluvial dunes 248
Fluvial process school 265
Fluvioglacial deposits 128, 222, 225
Foresets 256 ff
Formative discharge 266
Frontal storms 228
Froude number 248
Fulwood, Lancashire 60

Geochronological analysis 172, 175

Geomorphic thresholds 276
Gila River 3
Glacial burst 128, 239
Glacigenic diamict 253
Glen Feshie 186
Gley 193, 200
Gorges 240, 279
Grain pivoting 119
Gravel 189, 193, 222, 239, 242, 247, 253, 254, 257, 259, 260
Great Eggleshope Beck 125 ff
Growth curves 213, 215

Harthorpe Valley 220
Historical changes 220, 269
Historical sources 164, 228, 278
Holocene 187
Honiton, Devon 60
Howgill Fells 158, 159, 186
Huddersfield 58, 60
Hungary 53
Hurricane Charlie 158, 164
Hydraulic geometry 270
Hydrological models 5, 6, 11
Hyperconcentrated flow 161, 248

Iceland 239
Idaho 115
Imbrication 242, 278
Incision 166, 187, 194, 195
Indian canals 265
Induration 174, 179
Inter-arrival time 235
Inverse grading 242, 247
Iron 193, 195

Japan 24
Jokulhlaup 239, 248, 249

Katherine Gorge 172
Kenya 138, 144
Kinematic wave 52

Lag deposit 247
Lagged response 235
Lake District 1
Lake Estes 163
Lake Regina 128
Land use 185, 220, 233 ff
  change 29, 30, 57, 166
Landslides 209, 215, 273

Landslips 195, 207
Lateral accretion 280
Lateral-diagonal bar 258
Lawn Lake 158, 159, 163, 167
Leicester 60
Lichenometry 164
Lichens 185, 212
Lithofacies 255, 259
Little Ice Age 228
Llyn Eigiau 185
Lobate deposits 242
Loch Lomond Stadial 186, 194
Lowland rivers 214
Luxembourg 18
Lynmouth 152, 215

Macropore flow 18
Magnitude–frequency 3, 151, 161, 173, 179, 219, 220, 233, 239, 266, 278
Manganese 187, 193
Manning equation 84, 212
Maryland 61
Mass movement 24
Matrix flow 19
Matrix support 242, 245, 256
Maximum flood 215
Maximum likelihood 179
Mean annual flood 37, 213, 215
Meanders 207, 266, 268
Meltwater 186, 222, 229, 234, 242
 gorges 222, 229, 234, 235
Mining waste 164, 175
Miocene 130
Models 4, 5, 6, 11, 164, 176, 276
Momentum transfer 98
Monsoon 274
Moretonhampstead 60
Mosedale Beck 185 ff
Mount St Helens 158, 159

Nested thresholds 279
New Zealand 19
Newtonian behaviour 161
Non-cohesive gravity flow 159
Nonlinear dynamic systems 4
Non-Newtonian behaviour 159, 242
Noon Hill 158, 213

Okehampton 60
Openwork gravels 256
Ordovician 187

Organic drapes 174
Organic matter 175, 189, 195
Overbank
 deposition 164
 flow 96, 266
 storage 38
Overland flow
 infiltration excess 16
 saturation excess 22
Oxford 18

pH 195
Palaeochannels 171, 195, 207, 225, 268
Palaeofloods 107, 128, 171, 175, 178, 181, 215, 219, 248, 261, 268, 278
Palaeosol 174, 179, 189, 193
Palaeostage 171 ff, 178, 179
Palagonite 240, 245, 247
Paraglacial 235, 253
Partial Duration Series 266
Peat 158, 187, 202 ff, 229
 blocks 205
 slides 199, 202, 204, 215
Pebble stringers 245
Pedogenesis 193
Periodicity 229, 234
Persistence 273, 278
Phosphorus 195
Physical models 276
Pinhoe 60
Pipeflow 18
Planar cross-bedding 245
Planar stratified beds 256
Planform change 219, 225, 233
Podsols 187
Point-bar 207
Porlock 60
Porth Llwyd 185, 186, 187, 193, 194
Preston tube 93
Pseudobedding 247
Pseudoplastic 242
Pulsed flow 248
Pumice 240, 247
Pumice sands 242, 245, 248

Quasi-equilibrium 234
Quaternary 253

Radiocarbon dating 175, 249
Rafted boulders 242, 248
Rainsplash 16

Rating curves 84, 89, 116
Reactivation surfaces 257 ff
Recovery 155, 273, 276, 278
Recurrence interval 199, 213, 214, 215, 228, 233, 270, 273
Red River 58
Regime theory 265
Regional flooding 229
Regional growth curves 214, 274
Regionalization 180
Relaxation Time 4, 69, 233, 276
Relict channels 268
Resistance formulae 84, 212
Return period 199, 213, 214, 215, 228, 233, 270, 273
Reverse tangential grading 257, 259
Rilling 16
River
  Clyde 65
  Cree 30
  Dee 220, 225, 228, 234
  Ettrick 29
  Exe 36
  Feshie 220
  Gara 28
  Körös 53
  Rheidol 195
  Severn 253, 261
  Tees 38 ff
  Ure 136 ff
  Vyrnwy 270
River confluences 135, 270
River terraces 187
Roaring River 163, 167
Rotational slump 209
Roughness 84, 212
Rugby 58
Runoff 12 ff, 158, 229, 234

Salt River 181
Sand 164, 172, 189, 193, 253, 254, 255, 257, 259
  dunes 195
Sandur 239, 248
Savannah River, Georgia 27
Scour 144, 242
  and fill 162
  line 177
Secondary flows 91
Sediment
  concentration 159

Sediment (*cont.*)
  entrainment 108
  inputs 233
  -laden flow 159
  mobility 229, 234
  routing 4, 276
  storage 276
  transport 62, 96, 107 ff, 158 ff
  models 6
Selective entrainment 108 ff
Semi-arid climate 138, 144, 273, 274
Sensitivity 276
Shear plane 245
Shear stress 93, 109, 171, 278
Sheetwash 16
Shields diagram 126
Shields entrainment function 111
Silt 172, 193, 200, 239, 245, 247, 259
  drapes 258
  lines 172
Sinuosity 253, 268
Skelmersdale 60, 70, 73
Slackwater deposits 171 ff, 278
Slapton Wood Catchment 28
Sliding 119
Slope stability 273
Slope/area method 212, 215
Slurries 242
Snowmelt 155, 228
Soil
  compaction 13
  development 187, 193
  parent material 195
  profile truncation 195
  sequence 187
Solheimajökull 237
Solheimasandur 247, 249
Solute transport 25
Sorting coefficient 118
Souris Spillway, Saskatchewan 129
St Venant equations 52
Stage-discharge curves 84, 89, 116
Stationarity 164
Stevenage 60, 71
Storm runoff 12 ff
Stream power 269, 278
Subglacial 239
Subsurface stormflow 17
Suspended sediment 172, 273
Synsedimentary folding 255

Tephra 239
Tephrochronology 249
Terrace 186, 193, 212, 225, 234, 253
　incision 249
　remnants 195, 253
　slope 261
Thermoluminescence 175
Threshold of motion 111
Thresholds 4, 7, 161, 233, 236, 273, 279
Thunderstorms 200
Tractive force 260
Tree line 229
Tree rings 185
Tributary junctions 136, 172
Tropical environments 274
Trough cross-bedding 245

Unit hydrograph 214, 229
Universal soil loss equation 17
Unsteady flow 49, 89, 259

Upland catchments 214
Urbanization 57 ff

Valley storage 38
Variable source areas 12, 23, 24
Vegetation 204
Velocity 161, 278
Viscosity 159, 242, 248
Volcanic eruptions 187, 239

Water Resources Act, 1963 152
Water surface profiles 176
Weathering 193, 195, 249
Welded contacts 256
Widdale Beck, Yorkshire 136 ff
Wyoming 115

Yellow River 202, 203, 205, 209, 213, 214, 215
Yield strength 242, 248